Solutions for a Wounded Planet

JIM KINGHAM

One Printers Way
Altona, MB R0G 0B0
Canada

www.friesenpress.com

Copyright © 2022 by Jim Kingham
First Edition — 2022

All rights reserved.

No part of this publication may be reproduced in any form, or by any means, electronic or mechanical, including photocopying, recording, or any information browsing, storage, or retrieval system, without permission in writing from FriesenPress.

ISBN
978-1-03-913435-5 (Hardcover)
978-1-03-913434-8 (Paperback)
978-1-03-913436-2 (eBook)

1. *Nature, Environmental Conservation & Protection*

Distributed to the trade by The Ingram Book Company

DEDICATION

This book is dedicated to two remarkable young women, Malala Yousafzai and Greta Thunberg.

Malala demonstrated great courage in recovering from an assassination attempt and deciding to continue her battle for girls around the world to have an equal opportunity to get an education. She had been raised in a culture that denied that equal opportunity.

It was her determination to seek equal rights for girls that had led to the attempt on her life when she was just fifteen years old. Male assassins, believing that girls should not be educated, had shot Malala and two other schoolgirls. Malala was shot in the face.

Despite her suffering and disfiguration, Malala continued her battle for equal rights for girls, even when confronted with more assassination threats.

This book concludes that the single most important step to solve the world's environmental problems is the empowerment of women around the world. That empowerment begins with ensuring that girls have the same access to an education as their male cohorts, everywhere in the world. Women who are denied control of their own minds and bodies in male-dominated cultures are thereby denied the opportunity to contribute to the resolution of the world's most pressing environmental problems.

Greta Thunberg is a shining example of how young women, with the full benefit of an equal education and equal standing with men in the world, can contribute to solutions for a wounded planet.

Thank you, Greta and Malala, for showing the way.

ACKNOWLEDGEMENTS

Thanks, first, to Mom and Dad. Pearl and Ernie Kingham raised me in a natural environment on the outskirts of Kitchener, Ontario. I was encouraged to work and play in that environment, tending to our substantial vegetable garden, walking through the neighbour's woodlot, up to my knees in a muddy stream examining dragonfly nymphs and minnows, and so on. Under their firm but gentle care, I began to understand and respect the relationship of everything to everything else.

Thanks, too, to the many dedicated Kitchener-Waterloo teachers from kindergarten to graduate school who made lasting impressions. Most notable among them was high school teacher Mr. Arnold Boggs, who taught the beauty of global air circulation models and the mysteries of tectonic plate movements—long before these were formally in the curriculum. Thanks to another great teacher, Mr. Sykes, who inspired me to become a scientist.

I am grateful to all my colleagues in the environmental field who worked alongside me as we took those first tentative steps to protect the local, national, and global environment.

A special thanks to my hiking buddies, Don Armstrong, Roger Bailey, Chris Chapman, Klaus Gothe, Richard Lake, Michael Randerson, Clive Timms, Peter Vivian, Patrick Wesley, and Graham Williams for their always stimulating, and often challenging, discussions as we trekked through thousands of kilometers of magnificent old growth forests, and scaled a few mountains along the beautiful west coast of Canada. While we may have disagreed on some specifics, our conversations were always considerate, and I believe we all agree on the overall thrust of what needs to be done.

Thanks to my son, Scott, for his help with some of the diagrams, and a huge thank you to my beloved Penelope for her constant encouragement to keep at it 'til the job was done.

Foreword

BY ELIZABETH MAY

"We stand at a critical moment in Earth's history, a time when humanity must choose its future."
OPENING WORDS FROM THE EARTH CHARTER

We are at a time of great peril. The planet is wounded; the world is wounded. Healing is an urgent priority.

Into this situation of complex and interrelated threats – the climate emergency, the biodiversity crisis, social injustice and a pandemic – Jim Kingham has taken on a brave project. Although he admits in these pages that the task was difficult and one he would have preferred someone else had taken on, he has succeeded in making a serious contribution.

This book takes us through the science, the politics, even the moral and spiritual dimensions of the crisis and emphasizes solutions. We can make a difference.

One aspect of this book that is much needed is the history of environmental protection and Canada's role. It was in Jim's role as an Environment Canada scientist that we first met. As the Environment Minister's Senior Policy Advisor from 1986-88, I had the honour of contributing to solutions for the Great Lakes, to fight acid rain and to protect the ozone layer.

As the years pass and we fail, as a global community and within Canada, to effectively address the climate crisis, it is helpful to look back and ask why we succeeded then, but fail now?

Having had a front row seat to these events, I believe that, over time, our political class has become increasingly subservient to transnational corporations. Where former Prime Minister Brian Mulroney, or Ontario premier David Peterson, stared down big polluters – whether to fight acid rain or shut down ozone depleting chemicals - recent political leaders whether Justin Trudeau or BC's John Horgan seem to be incapable of listening to science and acting – whether to shut down greenhouse gas (GHG) emissions or protect old growth forests.

I think it is worth examining why the 1987 Montreal Protocol to protect the ozone layer worked and the 1997 Kyoto Protocol to act for climate – only ten years later – was ignored by the US and Canada and is viewed as a failure. True, the European nations largely met and exceeded their Kyoto targets, but the goal of reducing global emissions by roughly 5% was not met. In fact, GHG emissions have climbed globally in every year since we began negotiating climate action back in 1990 – right up until the COVID pandemic.

The horrible reality is that we have emitted more GHG between 1990 and today than between the beginning of the Industrial Revolution and when we started calling for reductions.

I think it is worth considering that back in 1987 when we negotiated the Montreal Protocol and 1997 and Kyoto, the same governments that negotiated those environmental pacts completed the Uruguay Round for world trade and set up a more extensive regime for global trade and corporate rights under the World Trade Organization. In 1987, governments made the Montreal Protocol effective by including trade sanctions as punishments for any government that ignored the treaty, by 1997 those same governments, including Canada, blocked having any sanction in the Kyoto Protocol. So too, the Copenhagen Accord and the more recent and comprehensive Paris Agreement has no enforcement mechanism other than group shaming and demonstrations in the streets. It is time to tame the power of global corporate rule by ensuring that trade agreements do not impede progress to protect a livable future.

We have to find shared solutions at all levels – the small things we can all do, while pushing governments at all levels to make the big decisions – ending subsidies to fossil fuels, canceling new fossil fuel infrastructure like the Trans Mountain Pipeline, banning fracking and treating fracked gas like coal – as they have the same carbon footprint. Both need to be phased out and fast.

Solutions include the cultural shift to a life less focused on material wealth and more on a life filled with meaning.

We have solutions – whether technological, financial, economic and scientific. But, like Jim, I think the real shift is in our spiritual perspective. As Pope Francis wrote in his remarkable encyclical, Laudato Si, we have to recognize in the intersecting crises of social inequity and environmental damage the role of greed, materialism and a "throw away culture."

The papal letter raises the issue of justice between the wealthy nations and the Global South: "We have to realize that a true ecological approach always becomes a social approach; it must integrate questions of justice in debates on the environment, so as to hear both the cry of the earth and the cry of the poor."

We face a profoundly moral crisis.

And we come full circle, back to the words of the Earth Charter as invoked by Pope Francis:

"The Earth Charter asked us to leave behind a period of self-destruction and make a new start, but we have not as yet developed a universal awareness needed to achieve this. Here, I would echo that courageous challenge: 'As never before in history, common destiny beckons us to seek a new beginning… Let ours be a time remembered for the awakening of a new reverence for life, the firm resolve to achieve sustainability, the quickening of the struggle for justice and peace, and the joyful celebration of life.'"

Table of Contents

FOREWORD BY ELIZABETH MAY VII

PREFACE XIII

THE CONCEPT XXI

PART I: THE TRAGEDY OF THE GLOBAL COMMONS 1
   Air 2
   Water 35
   Land 46
   Species 53

PART II: WASTED BY DESIGN 59
   Everywhere You Look 60
   Things That Are Hidden From View 64
   Food 70
   Shelter 78
   Security 82
   Health 89
   Energy 95
   Transportation 108
   Consumer Goods 115
   Recreation 121

PART III: SOLUTIONS FOR A WOUNDED PLANET 125
   Actions at the personal level 126
   Actions at the Family level 136
   Community Actions 142
   Municipal Initiatives 149
   Regional Actions 155
   Provincial Action 158
   National Initiatives 161
   International Initiatives 172

CONCLUSIONS 185

WHY I WROTE THIS BOOK 189

Preface

IMAGINE

Imagine a modest house on the outskirts of a medium-sized town in North America a hundred years from now. Several people are sitting around the kitchen table. They are all descendants of a grandmother who left a strange envelope in her will; an envelope to be opened by them in 2125.

The envelope contains a letter:

"Dearest Grandchildren,

A hundred years ago, when I was in my prime, the world was in terrible shape. The environment was being ruined, there were too many people, many beautiful and useful animals were going extinct, and millions of poor people were fleeing their homes in hopes of finding a richer life somewhere else.

The scientists told us what was happening. They spoke about the pollutants in the air and water, the destruction of our forests and oceans, the waste in our diets, and what would happen if other species were to die away.

I believed the scientists and did everything I could to help solve the problems we had created. I tried to convince all my family and friends, and my neighbours and politicians to do the right things. It was hard. It cost me a lot.

I don't have any money to leave you, but I hope I have left you a world that is better than it would have been if I hadn't tried my best.

Love to you all and keep up the good work. Grandma."

The readers of the letter looked out the window. Their world was not perfect but at least the air was cleaner than it had been back then; there were still fish in the oceans and bees in the meadows. Populations had stabilized and there was enough food for all, although diets were a lot different than they were back then. The grandchildren had read about the troubles of the 21st century and the debates about what needed to be done.

It looked like Grandma and her friends had succeeded—she was always such a persuasive and caring woman. Together they raised their voices:

"Thanks Grandma. Thank you, thank you, thank you!"

Now, imagine another modest house on the outskirts of a different medium-sized town in North America a hundred years hence. Here too, several people are sitting around the kitchen table. They are all descendants of a Grandfather who left a strange envelope in his will; an envelope to be opened in the presence of all his grandchildren in 2125.

The envelope contains a letter:

"Dearest Grandchildren,

A hundred years ago, when I was in my prime, the world was in pretty bad shape. The economy was being threatened by scientists who spoke wildly about pollutants in the air and water, the destruction of forests and oceans, overeating, and what would happen if other species were to die away.

Hogwash! I did everything I could to keep the economy on track. I tried to convince all my family and friends, and my neighbours and governments to do the right things and keep the economy ticking along.

I hope I succeeded. Here's $5,000 for each of you. Love Grandpa"

The readers of this letter looked out the window. Their world was quite miserable. The global temperature had gone up 5°C. They couldn't go out without an air filter mask. Sea levels had risen and now quadrillions of dollars were being spent each year to cope with flooded coastal cities and frequent storms and hurricanes. Drought and wildfires were annual events at the end of hurricane season. They spend more than half their income just to cope with environmental disasters and to maintain the "Stratospheric Parasol," that layer of sulphur particles shot into the upper atmosphere every few weeks to try to cool things down a bit—it was impossible to see the sun clearly anymore. Ocean fishing had been ended because so many species were fished out. There was hardly enough food for the rich world, with almost nothing left for the poor world. Global population has grown to 12 billion people, most of whom are desperate to be somewhere else.

The grandchildren had read about the troubles of the 21st century and the debates about what needed to be done. How did they feel now?

> "Oh Grandpa! How could you, how could you, how could you! The evidence of environmental damage was so clear, but you turned your back on it. Adding insult to injury, you persuaded your friends to do the same! And what is this $5,000 in old dollar bills? Your money has become practically worthless because the world economy crashed along with the environment. You can't buy a meal with 5,000 of those old dollars.
>
> Thanks a lot, Grandpa—thanks a lot!"

So, the question is:

What will *your* legacy be?

How do *you* want to be remembered?

It is in your power to provide for a better future. There are so many things that can be done by individuals, families, communities, industries, and countries to heal this wounded planet. Now is the time to start.

> *[...] the human suffering and cataclysms we are now experiencing need never have happened, [...] they occurred not through chance or the will of malevolent gods, [...] the revenge of nature and the devastation of our civilization are direct results of the uncaring arrogance of our forebears and of our own self-indulgent greed and neglect. What we have suffered is our own fault, and only through our own efforts can it be reversed and a hopeful and sustainable future secured.*
>
> Maurice Strong, in *Where on Earth are we Going?* (Knopf Canada, 2001)

In the introductory chapter of his book, Maurice Strong describes a hypothetical world in the year 2031 where anarchy and terror reign and the globe's environment is devastated. At the conclusion of that chapter, he describes a fictional charismatic spiritual leader, Tadi, who instructs the people with the words quoted above. I believe that quote applies today, without the global anarchy, but with enough cataclysms and more than enough suffering.

This book chronicles some of the missteps we are now making, in the hope that, by learning from these deliberate errors, we may correct our behaviour before it is too late. Our influence on planetary health was pretty insignificant for millennia, but it is terribly significant now. To put human existence in the context of life on this planet, imagine that you could make a stack of thousand-dollar bills twice as high as the Eiffel Tower. Each bill represents a thousand years. You can hardly see the top of the stack of bills half a kilometer above your head. Now take a handful of these bills, about 4½" thick, from the pile. That represents, approximately, the time humans have been on this planet. The last two bills in your hand represent the last two thousand years. Human beings had almost no effect on the planet for the time represented by all those thousand-dollar bills. But in the time represented by the last one-tenth of the last bill, human activity has had a catastrophic effect on many species of life, and our actions have begun to affect the basic equilibria of the entire planet.

THE AGE OF WILLFUL DESTRUCTION

I believe that we are in the third and most dangerous stage of human activity with respect to the global environment. For me, the four stages are the age of innocence, the age of ignorance, the age of willful destruction, and an age of global environmental enlightenment yet to come, in which we will re-establish and foster a new, sustainable, dynamic equilibrium between humans, other species, and our environment.

We have long understood that it is wrong to steal from another human being, but we were unaware, and thus innocent, of how wrong it is to steal limited natural resources from future generations and deliberately damage the global environment for immediate gain. That age of innocence is long past.

Somewhat later in our history, we humans had a sense that certain of our actions might damage the broader environment, but we lacked compelling data that showed the seriousness of that damage on a global scale. That age of ignorance has now also passed.

I believe that we are now in the third and most dangerous stage of human development in terms of our relationship to the global environment and our impact on that environment. This is the age of *willful* destruction—an age in which we know the damage we are doing. We now know how much of the Earth's non-renewable resources we waste. We know that things we dump on the land, discharge into the water, and emit into the air can have both immediate and far-reaching effects. We know that we are growing our population to a point where we will likely overshoot

Earth's carrying capacity and that too many people, and the animals on which they depend, already find themselves crammed into unlivable spaces. We know that religious practices established to help us learn how to coexist peacefully are being co-opted with dreadful effect, not only in terms of human suffering and degradation, but also in terms of the misuse of the planet's resources. And we uncritically accept the direction of our politicians, away from assisting with the provision and protection of those elements essential to our existence and towards the protection of vested interests that seek to garner ever larger shares of the limited supplies of "stuff" available on this planet. We are no longer ignorant of the damage of these actions, but we continue them. Hence the "Willful Destruction" term applicable to this age.

The fourth age, one of environmental enlightenment, lies ahead of us. The road is not easy, and there will be disasters along the way. The challenge is to get from where we are to where we ought to be with the least suffering and loss; not only for us humans, but for all species that depend upon our shared environment.

We find ourselves at the end of a geologic era called the Holocene Epoch. For much of that era, the earth enjoyed a dynamic equilibrium that involved the interactions of the many and varied living and non-living things on this planet. There was a relatively constant human population of perhaps one hundred million people, the alkalinity of the oceans was fairly steady, the oxygen, nitrogen, and carbon dioxide levels in the atmosphere were reasonably steady, and the global rates of phosphorus and nitrogen run-off from the land varied little. Some believe that we are at the beginning of a new age on the geologic time scale, the Anthropocene Era—an era in which humans disrupt the dynamic equilibrium that has existed for thousands of years as they tap ancient stored reserves of fossil fuels and other non-renewable materials for their present gratification. The unstable conditions that will ultimately flow from this human-induced disruption are beyond our calculation, but we know they will be bad.

EQUILIBRIUM

I will refer to the concept of equilibrium in general, and dynamic equilibrium in particular, from time to time in this book. The term is commonly used in chemistry. Here is a simple illustration of the principle. If you were to take a glass of water at room temperature and add salt to it, one teaspoon at a time, you would eventually reach a point at which the water could take up no more salt, and any more salt you added would settle to the bottom. That would be the equilibrium saturation concentration for salt in water at room temperature. It would be a dynamic equilibrium because, if you could see things at the molecular scale, you would see that, at the surface of the salt in the bottom of the glass, some molecules of salt would be dissolving while an equal number would be precipitating out of the solution, maintaining a constant equilibrium concentration of salt in the solution.

It is a dynamic equilibrium in more than just this sense, however, as it is also temperature dependent. Increasing the temperature moves more salt into the solution; decreasing it causes more to precipitate out. Planet earth has countless millions or billions of such dynamic equilibria. Until recently, it was inconceivable that a single species could materially affect the major, important equilibria of the planet. It is now quite clear that one species, man, has done just that.

It is not the reality that planet earth will find a new equilibrium point in the future that is troublesome—it is inevitable that it will do so. As an unconscious, unfeeling entity, the planet will not "care" what that equilibrium will be. What is troubling is the incalculable and unnecessary suffering that will be forced upon all life forms that share this planet as the new equilibrium is reached. Our generation and those that follow still have a chance to affect the path to the new equilibrium in such a way as to minimize such suffering.

I will describe the disequilibrium that lies ahead from the perspective of an environmental scientist and manager. My undergraduate studies of biology, chemistry, physics, and mathematics were followed by graduate research on the physical chemistry of biological systems. After graduation, I worked on various aspects of water pollution as a bench scientist.

I moved through the management levels of the Canadian Government's newly formed Environment Department, reaching the level of Regional Director-General for the Ontario Region, before moving on to the Provincial Environmental Assessment Board. Along the way, I helped draft the Canadian Ocean Dumping Bill to give effect to The Convention on the Prevention of Marine Pollution by Dumping of Wastes and Other Matter 1972, commonly called The London Convention. At about the same time I served on the Canadian delegation to the Law of the Sea conference as a scientific advisor on two specific areas of the draft convention, Marine Scientific Research and The Transfer of Marine Technology.

For a time, I was Director of the Canadian Government's Environmental Emergency Branch and head of the National Environmental Emergency Team, dealing with day-to-day emergencies and establishing an Environmental Emergency *Prevention* Program. The idea was that it would be better for the environment, and less costly, to prevent emergencies and thereby have fewer to respond to. The concept of preventing environmental problems at the local, national, and international levels rather than allowing them to happen and then trying to clean up afterwards, shaped much of the thinking presented in this book. As Canadian Chairman of the Great Lakes Water Quality Board, for example, I led a major initiative to identify and clean up toxic hot spots in the Great Lakes before they could destroy the water quality of the Great Lakes.

After decades with the Federal Environment, I was asked to become Vice Chairman of the Ontario Environmental Assessment Board, to adjudicate issues that could damage the environment, if they were badly executed.

Over my entire life, I have been gathering, analyzing, and organizing environmental information and knowledge. Now it is time to publish these observations in hopes that others will use this information and add their own observations and thoughts to heal this planet's wounded environment.

In this book, the term "environment" includes the usual air, water, and soil components, as well as broader concepts of resource use, economics, and cultural conditioning that affect the environment.

When tables or isolated numerical values are presented, they will reflect the situation up to the year 2021 when the manuscript was written. They incorporate my own work and work done with or by others, and will seldom be referenced because environmental variables are in constant flux. A given numerical value or table of values becomes dated as soon as it is printed because someone will have refined, corrected, updated, or revised the data upon which it was based. If you take issue with a given value and want to challenge, revise, or update it—great! I encourage you to challenge the data and gather more information. That alone expands your understanding of the scope of the problem and stimulates thinking about its implications. Nothing is static in this realm of inquiry.

This book is not intended to be another apocalyptic depiction of the end of the world. We are currently deliberately wasting the opportunity to arrive at a world in which a reasonably sized population might enjoy a high standard of living and a comfortable, sustainable environment. But there are solutions for the problems we have created. With implementation of these solutions, there is the possibility of a world in which everyone can live in equilibrium with the renewable resources of the planet and husband the non-renewable resources for their optimum use by present and future generations.

There was a time when one could argue that human actions that harmed the environment were the result of carelessness, ignorance, or stupidity. Falling back on such arguments is no longer acceptable. It is time to implement solutions.

There are plenty of "doom and gloom" publications that speak to global starvation, lack of drinking water, dwindling energy reserves, rising sea levels, expanding deserts, and global warming. While some or all of these effects may arise to some extent over time, the emphasis in this book will be on measures we can take to avoid the worst outcomes. While the prognosis may seem poor, it is possible that steps will be taken to prevent the more serious dangers ahead. It should even be possible to repair damage from past abuses of the planet as humans move towards a future of dynamic equilibrium on a healthy planet. Solutions to heal this wounded planet are found in Part III of this book.

We have gone from a relatively stable but dynamic equilibrium between the various components of the environment to an unstable disequilibrium. The position of the future equilibrium is still unknown but is not beyond mans' influence.

There are few who are still ignorant about the damage we are doing to our environment.

There are a few who know quite well the damage they are doing but want to continue that damage for personal gain. That must stop.

Most people know that we are damaging our environment. They may have a vague sense of unease. They may have definite views that something is wrong. They may have strong views that we need to do something about it. This book is intended to help those who want to leave a better environment to the next generation.

The Concept

This book is comprised of three parts:

    I. The Tragedy of the Global Commons

    II. Wasted by Design

    III. Solutions for a Wounded Planet

Part I builds on the work of William Forster Lloyd who, almost two hundred years ago, described how some people let their cattle overgraze common land in England. That led to the destruction of those lands, to the detriment of all who had previously shared the resource without trying to take more than their fair share. Garett Hardin expanded on this concept in his 1968 article in the journal *Science*. In this book, the concept is further expanded to assess the effect of human activity on the globally-shared environmental commons: the air, water, land, and species needed for the survival all people on earth.

Part II assesses human activities in order of their importance for survival: food, shelter, security, health, energy, transportation, durables, and recreation. It examines how human activities have changed over time in such a way that resources are now deliberately wasted in day-to-day activities. Often, people are left without reasonable options to act in ways that will not damage the environment, because things they need for survival have been designed to waste the environment, and they are offered few choices to avoid the waste. The scope of the text in this part is not limited to what people normally think of as waste; that is, things we dispose of when they are no longer useful to us. Rather, it is intended to be much broader. Here it applies to wasted opportunities, wasted lives, and wasted resources. In almost every aspect of human organization, we are deliberately pushed to ignore opportunities to conserve and enhance the environment.

Part III offers solutions by which this wounded environment may be healed. It is structured to move from solutions at the personal level, over which individuals have the most control, through to international solutions which could have the largest impact on protecting and repairing the environment. The top-level actions require the collective will of countries and the people they represent around the world. Where appropriate, the solutions proposed examine the activities described in Part II and suggest actions that may be taken at each level.

The time of reckoning has arrived. We must now learn how to live within current constraints if we are to enjoy the benefits of a healthy environment and a healthy economy; an economy that depends on that healthy physical environment.

PART I
THE TRAGEDY OF THE GLOBAL COMMONS

Almost two hundred years ago, William Forster Lloyd described how some people let their cattle overgraze common land in England. That led to the destruction of these lands, to the detriment of all who had previously shared the resource and benefitted from that sharing, when none took more than their fair share. Lloyd's example was that of a closed system, in which only those around the commons benefitted from its productivity, to the exclusion of outsiders. When individuals, acting in their own self-interest, over-used or contaminated the commons, all eventually suffered. Lloyd called this "The Tragedy of the Commons."

In this book, the concept is expanded to assess the effect of human activity on the broader commons, the air, water, land and species needed for the survival all people on earth—the global environmental commons.

This section of the book is presented in four chapters, Air, Water, Land and Species. This order was chosen because it parallels the order of importance of these environmental components: we can last but a few minutes without air, a few days without water, and a few weeks without food. How long we can survive when land and the variety of species, on which we depend for survival, are degraded beyond repair is still an open question.

Air

"This is the excellent foppery of the world, that, when we are sick in fortune—often the surfeit of our own behaviour—we make guilty of our disasters the sun, the moon, and the stars: as if we were villains by necessity; fools by heavenly compulsion; knaves, thieves, and treachers by spherical predominance; drunkards, liars, and adulterers, by an enforced obedience of planetary influence; and all that we are evil in, by a divine thrusting on: an admirable evasion of whoremaster man, to lay his goatish disposition to the charge of a star!"
(FROM SHAKESPEARE'S *KING LEAR*: ACT I, SCENE 2)

It is not surprising that our ancestors often blamed supernatural beings or distant stars for things that they could not understand. One of the earliest recorded instances of this is the belief of the ancient Greeks that the monsters Scylla and Charybdis were gulping and spitting out great quantities of water in the narrow Strait of Messina, between Italy and Sicily, and thus creating terrible currents and whirlpools that the Greeks had great difficulty transiting. Because they didn't understand how the gravitational pull of the moon could create tides and tidal currents at odd times, they ascribed their loss of life to mythical monsters that would suck them down into the depths. Subsequent knowledge led to an appreciation of when to avoid treacherous waters. Those who failed to do so were either ignorant, careless, or stupid.

When it comes to the protection of our most critical resource, the air we breathe, we can no longer claim to be ignorant. In this case, carelessness and stupidity appear to be related more to the selfishness of present generations than to any other factor. While we may take steps to protect the air in our most immediate environment, when it comes to global air quality problems, like atmospheric greenhouse gas control, we "make guilty of our disasters the sun, the moon and the stars" and "lay our goatish disposition to the charge of a star." In other words, we look for a scapegoat.

We exist because our home planet has been in a "zone of habitability" around our sun for the past three billion years or so. Under normal circumstances we could expect to enjoy this pleasant warmth for another one and a half to two billion years. But we risk moving out of the comfort zone sooner than we should because of collective human activity. To appreciate the "zone of habitability" concept, think of a mid-winter campfire next to which you are sitting with others. It is a bitterly cold night, and you would freeze to death without the warmth of the fire. So you find the ideal "zone of comfort" around the campfire. The people sitting around the campfire may be comfortable enough to remove some heavy winter clothing and still be comfortable. If they add more clothing, they risk getting uncomfortably hot. Overheating because of too much insulating clothing could become more than uncomfortable—it could become fatal. As we discharge energy-absorbing materials into Earth's atmosphere, it too will become more than uncomfortable; many life forms on which we depend will be destroyed.

CLIMATE CHANGE

Climate change is but the latest insult mankind has inflicted on the air we breathe. The first noticeable impact was likely the concentrated stink of human excrement. Humans learned the advantage of separating their cesspools from their living quarters. With the mastery of fire, they suffered particulates from smoke in their living spaces. So, they learned how to exhaust smoke from their immediate environment. Noxious hydrocarbon fragments in homes and factories were problems of a much later era. Humans expanded their environmental impact with the emission of toxic hydrocarbons into the outdoor environment—the London smog of the mid-twentieth century was the epitome of bad behaviour towards the atmosphere at that time. Next, persistent toxic chlorinated hydrocarbons were released into the atmosphere in such a manner as to allow their spread around the world. Then compounds that would destroy the ozone layer, which for billions of years had intercepted hazardous radiation from outer space, were released. Let's examine each of these deliberate insults in more detail and then return to the global warming issue, which is now the greatest atmospheric threat.

There are many kinds of air pollution: light pollution, sound pollution, toxic compounds, particulate matter, substances that affect the function of the atmosphere as a radiation shield, and molecules that change the way in which the atmosphere blankets the earth to prevent it from cooling down. Sometimes different pollutants have more than one of these effects, as in the case of particulate matter that may contain both toxic compounds and molecules that affect our radiation shield and the warmth of our atmospheric blanket.

LIGHT POLLUTION

People seldom think about light pollution and, if they think about it at all, they think about it mostly in terms of its interference with astronomy. But light pollution also disturbs peoples' sleep and disrupts ecosystems, particularly the component of ecosystems that involve birds that are active at night. Bird navigation is often disrupted, and each year millions of birds die as a result. The predator-prey relationship of many species (think of spiders and moths) is also affected. Excessive indoor light can also have negative effects on human health because of both its quantity and quality. Unfiltered halogen light can be harmful to the skin and eyes and many people are uncomfortable spending much time under fluorescent light.

In the context of this book, excessive lighting is a simple but clear illustration of how our resources are being wasted by design. Driving along a main road in almost any city in the middle of the night reveals a dazzling display of wasted light. Bright light advertisements, millions of streetlamps, and building exteriors lit at night, clearly illustrate the waste. Seasonal holiday lighting displays add another significant component of waste light and energy. The associated waste of non-renewable resources and the volume of pollutant by-products is incalculable.

NOISE POLLUTION

In the medical world the terms "acute" and "chronic" pain are used to distinguish between brief episodes of often quite sharp pain in contrast with more or less continuous pain. Noise pollution could be categorized this way as well. The intense and extreme noise of a high-powered motorboat is an example of acute noise pollution. Far more pervasive, and perhaps more serious, is the on-going hum of human activity, the noise of machinery and masses of people; noise that crowds out the sounds of nature that have been a source of survival information and comfort for most of human history.

We are learning to live without silence or the sounds of nature. How often do we see people with "ear buds" in their ears, plugged into some sort of playback device, whilst engaged in an otherwise benign natural activity like gardening or jogging through the park? While this may not be a significant waste of the acoustic environment. it may prove harmful to the listener in many other ways.

Sound is a form of energy. It takes energy to produce. While noise is not a deliberate waste of resources in and of itself, it is often an indicator that energy is being wasted. The unmuffled racket of cars, motorcycles, and speedboats are but a few examples of such waste. The designation of certain areas for the quiet enjoyment of all would have an ancillary benefit of at least not wasting resources to produce noise in those areas. We would be wise to challenge those who create disruptive noise for frivolous purposes, and to put in place appropriate controls to prevent such deliberate waste.

TOXIC COMPOUNDS

I imagine our ancient ancestors avoided toxins quite naturally. As we evolved, we learned to stay away from the sulfurous fumes of volcanoes, avoid the odious odours of fetid swamplands, give a wide berth to decomposing corpses of animals, and separate ourselves from the smell of our own excrement. These smells were repugnant because they were associated with either outright toxicity or the transmission of disease.

We are now beset by quite a different set of airborne toxins—substances that may have no odour, but that could have immediate or long-term deleterious health effects. Some are immediate and local, others long-term and pervasive. Some are now well characterized; others are, at best, questionable and, at worst, unknown.

Cigarette smoke is an example of an airborne toxin which has now been well characterized. We know that it can sicken or kill smokers and injure innocent bystanders who are unfortunate enough to be unable to avoid it. Despite this knowledge, many people around the world still deliberately poison their local air environment with cigarette smoke. As we educate ourselves about the toxicity of this pollutant and sales in certain places drop, those who manufacture cigarettes deliberately move their operations abroad to profit by poisoning distant atmospheres. It is not an accident, nor is it ignorance, but rather a matter of turning a profit wherever possible, regardless of the resultant suffering. In this example, the deliberate waste is far more pervasive than is immediately obvious. Valuable land, water, and fertilizer are wasted to produce tobacco, a product designed to kill. Additional resources are unnecessarily consumed to treat those sickened by cigarette smoking.

Now let's take this aspect of "deliberate damage to the atmosphere" a step further. We have learned that asbestos fibres cause lung cancer and other respiratory diseases.

Unlike cigarette smoke, these tiny particles have no odour and often cannot be seen. Many parts of the world have banned the use of asbestos because of its deleterious health effects and insidious nature. In other parts of the world, its use is still permitted despite knowledge of its negative effects. But some countries in which its use is banned still export it to jurisdictions in which it is allowed. Canada is the primary culprit here. As recently as 2011, Canadian Prime Minister, Stephen Harper said: "Canada is one of a number of exporters of chrysotile (a form of asbestos) and there are a number of countries, in which it is legal, who are buyers. This government will not put Canadian industry in a position where it is discriminated against in a market where it is permitted." He said this even though Canada's leading medical authorities told him that the export of asbestos was medically and morally indefensible and would lead to an epidemic of asbestos-related disease and death overseas.

The World Health Organization (WHO) estimates that as many as 107,000 people die each year from asbestos-related illness. For a third time the Canadian government alone stepped forth to prevent the inclusion of asbestos on a list of exports requiring warnings of health hazards to recipient countries. With such warnings those countries could then refuse asbestos imports if they believed they were unable to handle the product safely. This is another clear example of how air quality is wasted by design for immediate economic gain.

There are hundreds, if not thousands, of such hazardous substances emitted around the world. The United States Environmental Protection Agency lists about two hundred known airborne toxic substances or compounds. If we know them and can list them, but do not control them, we are deliberately polluting our atmosphere.

PARTICULATE MATTER

Very small particles discharged into the air are called Atmospheric Particulate Matter; particles so small that they can be buoyed up by the motion of air molecules. Air molecules move about in the atmosphere in proportion to their temperature; warmer molecules move faster and cooler ones move slower. If particles are small enough and they are struck by an upward-moving air molecule, they will be bounced upwards and be kept in the air by the jostling of air molecules, just as a volleyball can be kept up in the air as long as players can get under the ball and give it one boost after another. Eventually someone will miss, and the ball will fall. Eventually particulate matter is not struck by enough upward-moving air particles, and it falls to the ground.

Particulate matter in the atmosphere is often what makes beautiful sunsets. It can be natural and benign, like sea spray, the biggest source of particulate matter in the atmosphere, or smoke from forest fires or grass fires. It can be man-made (*anthropogenic*) and quite harmful, like smoke from the burning of fossil fuels and other particulates from industrial processes. Natural sources account for about 90% of the particulates in air. The 10% of atmospheric particulates from human activity are linked to health hazards such as heart disease and lung damage.

Each year, a typical coal-burning power plant in North America generates about nine and a half million kilograms of particulate matter, which includes about:

- 50 kilograms (kg) of lead
- 77 kg of mercury
- 100 kg of Volatile Organic Carbon compounds (VOCs, which often react with other substances in the atmosphere to produce ozone)
- 102 kg of arsenic
- more than 450,000 kg of very small airborne particles, which can cause chronic bronchitis, aggravated asthma, and premature death
- more than 9,000,000 kg of sulfur dioxide (SO_2), which forms small airborne particles that penetrate deep into the lungs

Each such power plant also adds:

- more than 650,000 kilograms of carbon monoxide
- 9,000,000 kilograms of nitrogen oxides (NO_x)
- 3,000,000,000 kilograms of carbon dioxide each year.

When you consider that there are more than 670 coal-burning power plants in North America alone (646 in the USA, 22 in Canada, and 3 in Mexico), you can appreciate the volumes of toxic particulates released to the atmosphere each year. 670 coal-burning power plants would produce about 33,500 kilograms of lead, 51,600 kilograms of mercury, 67,000 kilograms of volatile organic compounds, 68,300 kilograms of arsenic, 300,000,000 kilograms of small particulates and 6,000,000,000 kilograms of sulphur dioxide, for a total of more than 6,300,200,000 kilograms of particulates thrown into the atmosphere each year from North American coal-powered energy production alone. I have deliberately included all the zeros here to emphasize the enormity of a single-word term like "six *billion* kilograms of particulates a year." This is just from North America. Adding in China and the rest of the world would likely triple the extent to which the air we breathe is deliberately contaminated with particulate material each year (coal use in China has increased five-fold to about 3,000,000,000 tonnes since 1985).

High-flying aircraft are also significant contributors to ground-level particulate pollution that has serious effects on human health. It is estimated that global emissions from aircraft cruising at an altitude of about 35,000 feet cause about 8,000 premature deaths each year. This is because the Nitrogen Oxides (NO_x) and Sulphur Oxides (SO_x) they emit combine with other gases already in the atmosphere to create very fine particles that are especially dangerous to human health. When the European Union proposed charging a fee for the disposal of these pollutants into the atmosphere, a fee that would amount to about $3.50 per passenger per trip, countries such as Canada, the United States, and China vigorously protested. Although the fee was intended to discourage air travel because of increased cost, it is so ridiculously small that it would undoubtedly have had no effect on the number of trips and thus would have done nothing to improve human health. It is astonishing that China should be opposed to these particulate fees since about 3,500 of those annual 8,000 excess deaths likely occur in China.

In the most crass interpretation of this opposition, China is essentially saying that it is not worth $3.50 per passenger to take steps to dissuade the airline industry from burning fuel inefficiently and thus help save the lives of some 3,500 of its citizens each year. It is doubly sad to observe that the 3,500 victims are not the airplane passengers.

This resistance to taking measures to encourage the protection of the atmosphere from aircraft emissions is another clear example of how politics and corporate imperatives work together to deliberately destroy a component of our environmental commons. The European Union initiative was intended to price in the costs to society of using fossil fuels. That action was expected to lead to reduced consumption of aviation fuels because of market pressures (people would presumably travel less because of higher costs). It was also intended to generate funds that could be used to help offset the costs of treating people made sick by inhaling these pollutants, among other things. The airline industry vigorously opposed such action because it would affect their bottom line. They had no concerns about the health or other impacts of their industry.

At my undergraduate university, McMaster University in Hamilton, Ontario, morning light often revealed a fine layer of reddish dust on vehicles that had been parked on campus overnight. An overnight east wind brought iron oxide dust from the smokestacks of the steel mills on Burlington Bay, five kilometers to the east. Decades later, I visited those same steel mills as Director-General of Environment Canada's Ontario operations; I was told point blank by the president of the steel company: "You can't make steel without making a mess." My years of work on environmental issues taught me that a more accurate summary would have been: "You can't make steel without making a mess unless you are prepared to pay to clean up your mess, and that will hurt the bottom line." Clearly the steel industry will deliberately contaminate the air and water in its vicinity to make a profit, as long as no one requires them to do otherwise. As far as the "bottom line" is concerned, one has to ask: "Whose bottom line?" If contamination causes health effects that have to be paid for by others, whether individuals or states, then the industry is improving its bottom line at the expense of others. If the steel industry is not installing the necessary bag filters and other paraphernalia needed to keep the air clean, then it is hurting the bottom line of the industries that make and sell such paraphernalia. It is just a question of whether society as a whole is prepared to demand that polluters pay the cost of polluting, instead of passing the costs of environmental damage on to others.

Under the auspices of the United States Environmental Protection Agency, the EPA, a Cross-State Air Pollution Rule (CSAPR) was developed to reduce Sulphur-dioxide emissions by 27% and nitrogen-dioxide levels by 46% from their 2005 levels. The EPA reckoned that would save 13,000 to 34,000 lives over the next three years or so and save between $120 and $280 billion in health care and other environmental costs. Of course, by October 2011 at least thirty-six parties, (state power companies, trade associations, cities, etc.) had appealed to US Courts to stop CSAPR. Their primary reason: it would have a negative effect on the economy. Here again, they really mean a negative effect on their own bottom line. The bottom line of those who produce equipment for pollution prevention is not their concern, nor is the huge societal cost of treating those sickened by such pollution. As long as we accept the uncontrolled release of particulates into the air, we are knowingly wasting the world's atmospheric resources.

RADIATION-SHIELD DESTROYERS

The Earth is constantly exposed to large doses of radiation from deep space, mostly from the sun. A good portion of this radiation is ultraviolet radiation. Excessive exposure to one form of this radiation, UV-B, can cause genetic damage, resulting in problems such as skin cancer. Fortunately, we have a shield against UV-B radiation in the form of ozone molecules in a thin layer of the atmosphere about twelve to eighteen kilometers above the earth's surface. This ozone layer is largely responsible for reducing the amount of UV-B ultraviolet radiation at the earth's surface to about 1/350 millionth of the amount above the ozone layer. Ozone (O_3) is three atoms of Oxygen linked together in a single molecule and is quite different from the normal

Oxygen atom, O_2, which we breathe. These two types of Oxygen molecules are very different from a chemical point of view as well.

Ozone is a gas that occurs naturally in the upper atmosphere. It is also produced at lower levels of the atmosphere by emissions from internal combustion engines and other human activity. In the air we breathe, ozone can be quite harmful to human health, causing shortness of breath, pain during deep breaths, wheezing, and coughing. It has been linked to increased hospital admissions and emergency room visits for asthma and other respiratory problems. It can also reduce the body's resistance to infection. Long-term, repeated exposure to high levels of ozone may lead to large reductions in lung function, inflammation of the lung lining, and more frequent and severe respiratory discomfort.

So, we should be doing everything we can to prevent ozone production at the earth's surface, while at the same time protecting it high up in the atmosphere to be our shield against UV-B radiation.

Unfortunately, ozone depletion occurs in many places in the Earth's ozone layer, most severely in polar regions, because human-produced trace gases that contain chlorine and bromine have been drawn up by updrafts over populated areas and carried at high altitudes to the polar regions where they interact with ozone and destroy it. Older refrigerants were a big problem; they were made mostly of chlorofluorocarbons and bromofluorocarbons, which are very stable compounds. Chlorofluorocarbons (CFCs) are stable, unreactive and non-toxic chemicals, which makes them ideal for use in manufacturing processes. They decompose slowly in the lower atmosphere and rise upwards into the stratosphere where they are transported by high-altitude winds to polar regions. There, they break down and release chlorine atoms that destroy the ozone.

It takes about ten years for CFCs to reach the stratosphere after they are released by human activity. In the stratosphere, each CFC molecule is responsible for the destruction of about 100,000 ozone molecules, over approximately one hundred years, before its constituents dissipate. Ozone has been destroyed at the rate of about 4% per decade in the northern hemisphere, with much greater destruction in polar regions where "ozone holes" have been created.

To get a sense of how deliberate mankind has been in this kind of environmental destruction, it is informative to review the events between the discovery of the damage potential of CFCs, in the early 1970s, and the acceptance of the Montreal Protocol to control CFCs decades later.

Dr. Rowland and his Post-Doctoral student Mario Molina of the University of California at Irvine published the results of their research on CFCs in a very prestigious, peer-reviewed scientific journal, *Nature*, in 1974. Industries that manufactured CFCs, like the DuPont chemical corporation, fought the CFC ozone depletion theory, regarding it as "science fiction" and "utter nonsense." The president of Precision Valve Corporation, which made aerosol spray-can valves, is said to have written to the Chancellor of UC Irvine to complain about Rowland's public statements. It is not unreasonable to conclude that such opposition was more directed to the preservation of the corporate bottom line than the preservation of the atmosphere.

Dr. Rowland and Mario Molina testified before the United States House of Representatives in December 1974, as a result of which new research programs were funded to clarify the CFC-ozone issue. Rowland and Molina's findings were corroborated, and the US National Academy of Sciences released its own findings on the matter in 1976. For his work on CFCs and their likely effect on the ozone in the atmosphere, Professor Rowland became co-recipient of the 1995 Nobel Prize in Chemistry, along with his former post-doc Mario Molina and another atmospheric chemist, Paul J. Crutzen.

The Vienna Convention and the Montreal Protocol have now addressed the ozone layer depletion problem by setting emission limits and timetables for compliance. Twenty-five years after the agreement was signed, atmospheric concentrations of most ozone-depleting compounds have been drastically reduced and the thinning of the ozone layer seems to be stabilized. It is expected that the ozone layer will recover in the coming decades.

I have taken the time to explain this sequence of events in some detail because it is the clearest single event I have witnessed over thirty years of work in the environmental field where powerful industrial lobbies have denigrated serious scientific research, to the point of ridiculing respectable scientists, in order to protect economic interests at the expense of our common environment. This "economics" aspect of deliberate environmental destruction is now the dominant driving force against actions to protect the environment, as we shall see in later chapters.

GLOBAL WARMING

This is currently the hottest topic in the environmental field. It is a fiendishly difficult topic, but fortunately it can be explained in straightforward terms. I'll begin with the clearest possible description of the subject and touch on some of the more complicated aspects later in this section.

When you go to bed on a cold night you pull on a warm blanket to keep warm. Your body is generating heat all the time and the blanket reduces the amount of heat lost. The atmosphere is the Earth's blanket; it keeps the Earth warm. If you pull on too thick a blanket, your body gets too warm. If Earth's blanket is altered by the waste we pour into it, the planet gets too warm. If you put a heat radiator in your bedroom you don't need as thick a blanket to stay warm. The sun is the equivalent of the "radiator in the room" as far as the Earth is concerned.

That, in a nutshell, is how the Earth's heat balance works. The Earth has a lot of internal heat, which manifests itself as volcanoes and hot springs, and it picks up more heat each day from the sun. The Earth's atmospheric blanket keeps planetary heat loss to an acceptable level so that we don't freeze to death.

Global warming is not a recently discovered problem. When I was Regional Director General for Environment Canada in Ontario, one of my first visits was to the headquarters of the Atmospheric Environment Service (AES) in 1982. AES scientists described the importance of data coming from a facility on top of a mountain in Hawaii, the Mauna Loa observatory. It had started measuring atmospheric CO_2 in the late 1950s. CO_2 levels rose steadily until, in 1982, they were 20 parts per million above their 1950s level. AES scientists knew, even then, that this would have serious problems for the world's future climate and economy.

American politicians were also being informed of the looming environmental risk. Unfortunately, the fossil fuels industry was first in the field to take action. Their action was to deny the science and persuade the politicians, who depended on them and their financial support, to not act. This was the same approach they had used with respect to chlorofluorocarbons and the damage to the Ozone layer. Successive governments continued to subsidize the fossil fuels industry to the extent of tens of billions of dollars while the problem of disposal of CO_2 and other greenhouse gases into the atmosphere just kept on growing.

Greenhouse gases are gases that make the atmospheric blanket warmer. While most of the atmosphere is Nitrogen, N_2, (78%) and Oxygen, O_2, (21%), the remaining 1% is critical. Unlike N_2 and O_2, which have only two atoms in each molecule, Carbon Dioxide, CO_2, has three atoms, as is the case for water vapour, H_2O, and Ozone, O_3. Methane, CH_4, has five atoms. These three-atom and five-atom molecules are the main "greenhouse gases" because their polyatomic nature allows them to absorb longer wavelengths of light, getting warmer as they do so. Think of putting a cup of water under an infrared light. It gets warmer because its molecules absorb the long wavelengths of the infrared light. Human activity is increasing the Carbon Dioxide, Methane, Ozone, and water vapour in the atmosphere every year.

I recently built a greenhouse using the same principles as the global greenhouse effect so that I wouldn't have to add heat to keep it warm. It consists of windows facing the sun, a masonry wall painted flat black on the interior shaded side of the greenhouse, and a dark masonry floor. The short to medium wavelengths of sunlight pass through the glass and warm the surfaces inside, just as short to medium wavelengths of sunlight pass through the atmosphere to warm the Earth each day. Having dark surfaces inside the greenhouse helps with heat absorption, just as forests, oceans, and pavement on Earth absorb more sunlight than snow-covered fields and ice-covered oceans. The important thing is that as the warm surfaces in the greenhouse cool down overnight, they radiate their heat as long wavelength energy that does not easily pass through the greenhouse glass. Similarly, the Earth's long wavelength energy is prevented from escaping the atmosphere because of "greenhouse gases." The following diagrams illustrate the similarity of the two cases. On the left, the short wavelengths get through to the earth's surface. They are radiated back as long wavelengths, some of which are blocked by greenhouse gases like CO_2 and CH_4. On the right, the short wavelengths pass through the glass into the greenhouse and the long wavelengths are blocked by the glass and don't get out. Both the earth and the greenhouse warm up.

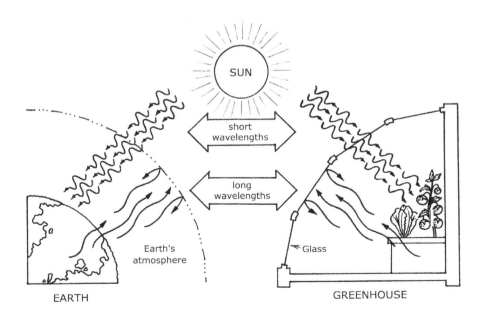

Figure 1. The Greenhouse Effect

The easiest way to appreciate the power of the greenhouse effect is to get in a car that has had its windows closed on a sunny day. The car will be quite hot because the shorter wavelengths of the sun's visible light energy pass through the glass and warm the interior surfaces. Heat builds up in the car because the warm surfaces inside radiate energy at longer wavelengths, wavelengths that do not pass through the glass windows easily. Similarly, long-wavelength energy from the earth's surface doesn't pass through all the CO_2 and other greenhouse gases in the atmosphere, and the planet warms up.

Humans and other animals contribute to the concentration of greenhouse gases in the atmosphere in different ways. We exhale carbon dioxide and water vapour. Plants recycle it back into carbohydrates for our consumption—a beautiful natural equilibrium. Humans have destroyed that equilibrium by unlocking the chemical energy stored in carbon reserves over billions of years, releasing heat and CO_2 in the process. The natural equilibrium of plants storing the carbon dioxide that animals exhaled, removing it from the atmosphere and putting it into beds of coal and pools of oil, has been reversed at a rate that is tens of millions of times faster. Carbon baked out of Carbonate rocks to make cement is another significant contributor to this disequilibrium.

There are more than 120,000 billion metric tonnes of Carbon in the Earth's surface and atmosphere combined (1 metric tonne = 1.1 US tons). 65,500 billion tonnes of that Carbon is stored in rocks near the surface, and 10 to 15 times that amount is in the rocky interior of the planet. Nearly half of the rest is in deep ocean sediments. Here's the total available carbon situation as best we know it.

Plants	550
Atmosphere	800
Oceans	1,000
Soil Carbon	2,300
Reactive Ocean Sediments	3,000
Fossil Carbon	10,000
Deep Ocean Sediments	37,000
Near Surface Rocks	65,000
TOTAL	120,500 Billion Tonnes

The critical part of carbon redistribution is the much-accelerated movement of carbon from the fossil carbon compartment to the atmosphere. At the same time, humans are removing large forested areas from the pool of forests that could store some carbon for longer periods of time and replacing them with plants that release it sooner. The transformation of carbonate rocks into cement also releases large quantities of carbon into the air.

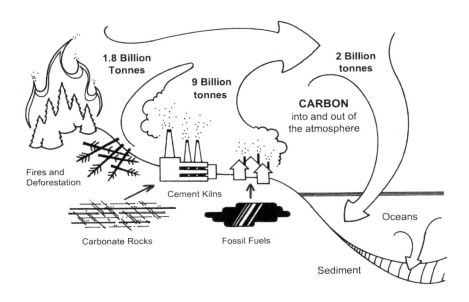

Figure 2. Carbon moving from the ground into the atmosphere and oceans.

Until the last two centuries, humans were unable to influence global climate. Our ancient ancestors ejected very little material into the atmosphere—certainly not significant in comparison with natural phenomena. Their settlements and deforestation were also puny when set against the effects of volcanoes, and so on. Even the onset of human agricultural activity had little effect. Forests were cleared and fields plowed but the combined effects were still too small to have a discernable impact on climate. The industrial revolution, on the other hand, had measurable effects on the world's climate. Some of these effects are cooling effects and others are warming effects. Overall, the warming effects more than offset the cooling effects, resulting in global warming. Human activities add about 10.8 billion tonnes of carbon to the atmosphere each year. The oceans take out about 2 billion tonnes, leaving a net increase of almost 9 billion tonnes a year. Because this is in the form of Carbon Dioxide (CO_2) which is 3.66 times heavier than Carbon, the net increase of CO_2 from human activity is about 36 billion tonnes

Now let's try to visualize the annual volume in global warming gases dumped into the atmosphere by an average Canadian family:

Imagine that your neighbour stops by in his pickup truck at 7:00 AM one day and drops off half a ton of horse manure in your front yard. Ten minutes later he comes back and dumps another load. He does this every day throughout the year. At the end of the year, the accumulated mess is a mound about 7 metres high (22 feet) and about 61 metres across (200 feet) at the base. It smells—it's ugly. It's not exactly an environmental disaster right now, but the rainwater percolating through it year after year will eventually contaminate the groundwater below. A new year begins; your neighbour does the same thing, day after day, year after year.

You might think the neighbour should dispose of his manure somewhere else, and he should pay to have it properly disposed of. But then, you do the same thing to him. Day by day, year by year.

The volume of CO_2 the average Canadian family dumps into the atmosphere each year is about 27,200 cubic meters. Enough to make a mound the same size as that mound of manure. You don't see it. You don't smell it. But it is there. It will eventually damage the atmosphere beyond repair; but you don't want to pay for your disposal and neither does your neighbour.

Collectively, humans dump about 20 trillion cubic meters of CO_2 into the atmosphere every year. The collective mound, if it could be seen dumped on the ground, would be about 54 kilometers (33 miles) across and about 13 kilometers (8 miles) high. Now that would be a huge pile of manure! The problem is, we don't see this CO_2 pollution. We don't smell it. Collectively, we seem to want to avoid being charged for the damage we do. But the CO_2 does its damage; year after year it builds up and the world heats up. Would we adopt a different attitude if the CO_2 that damages our atmosphere was a black, acrid, foul-smelling gas? I suspect so.

To encourage us to ignore our misbehaviour, many of those who profit from uncontrolled carbon emissions try to convince us that we are doing no harm. They choose to deny that human activity is causing climate change and thus continue along a path of human behaviour that is at least as old as Shakespeare's writings, which I used to begin this chapter. We would rather blame climate change on the sun than take responsibility for the mess we have made. When things go wrong, the anti-environmental industrial complex searches for a scapegoat. In this case, the blame is directed towards our own star, the sun. Thus: "It's just the normal variation in solar activity that is causing climate change. It's not us!" They choose to not even acknowledge that the "change" in climate change, is global warming.

FEEDBACK LOOPS

If the only effect of our collective misbehaviour in this area was the addition of CO_2 to the atmosphere, we would probably survive the ill effects. Indeed, those who deny the "human" aspect of "human induced global warming" delight in pointing out that the so-called *atmospheric forcing* which arises from our addition of CO_2 alone is quite small. Unfortunately, the effect of the CO_2 and other greenhouse gases we emit is multiplied by phenomena called positive feedback loops. "Positive" in this sense, does not imply something nice. Rather, it means something that magnifies itself until it becomes unbearable.

The concept of positive feedback arose in the 1920s and was primarily related to amplifiers. That is still the easiest way to understand the effect. If you were in a hall when someone switched on a microphone in front of speakers to which the microphone was connected, you would probably experience an excruciating screech emanating from the speakers. That screech quickly becomes unbearable, and the microphone or amplifier is switched off. The way it works is that the microphone picks up a small sound, amplifies it, and sends it to the speakers. The speakers project the amplified sound to the microphone, which picks it up and sends it back to the amplifier as a much louder sound. The amplifier then amplifies this louder sound and this positive feedback loop quickly grows out of control.

In the natural environment there are many positive feedback loop possibilities. They occur when a natural dynamic equilibrium is disturbed. The relationship between plants and animals is one such natural dynamic equilibrium that had stabilized over millions of years. As plants grow, they take in CO_2 and water and produce oxygen (O_2) and carbohydrates. Animals breathe in the oxygen, eat the carbohydrates, and produce CO_2 and water. Excess carbon was sequestered in the form of coal, oil, and natural gas, and was buried by sediments. Unfortunately, over the last century we have extracted that stored carbon and added it to the environment in the form of CO_2 and other greenhouse gases. The equilibrium has been disrupted and the positive feedback loops are beginning. Just like the screeching amplifier, these loops will inevitably produce unbearable effects.

The most obvious positive feedback loop has to do with what scientists refer to as the earth's *albedo*. Simply put, it is how reflective the surface of the earth is: the more reflective, the more sunlight is sent back into space; the less reflective, the more of the sun's energy is absorbed by the earth. The effect of changes in the albedo is most marked in the Arctic. This part of the world used to be covered with snow and ice for a large part of the year. Every spring, as the North Pole began to tilt towards the sun, the snow and ice began to melt, and the Arctic Ocean and the surrounding circumpolar lands could begin to warm up. In autumn the temperature would drop, the oceans would refreeze, and snow would again cover the land.

Over the last few decades, as the air over the Artic got warmer, more snow and ice melted. The water in the Arctic Ocean got a bit warmer because it was exposed to the sun for a longer period. When autumn arrived, it took longer for the surface of the Arctic Ocean to freeze because it took longer for the surface waters to cool down—the water and air above it were warmer than before. The circumpolar lands (Canadian, Russian, and Scandinavian) got warmer because their dark surfaces were exposed to the sun for a longer period of time. It took longer for fresh snowfall to cover the land without melting. On June 20, 2020, the town of Verkhoyansk, one of the coldest inhabited places on Earth, reached a summer temperature of 38°C, the hottest temperature ever recorded in the Arctic.

This is where the positive feedback comes into play. The longer the oceans and circumpolar lands are uncovered, the more of the sun's energy they absorb. The more of the sun's energy they absorb, the longer they are uncovered. Each year, on average, the situation gets a bit worse. Eventually the entire Arctic Ocean will be exposed at the end of the summer and the Earth's albedo will be significantly changed.

As is often the case with the dynamic equilibria that have kept the Earth in balance for millennia, a shift in one equilibrium affects others. This is especially true in Arctic areas. First, the equilibrium between stored carbon (coal, gas, and oil, for example) and CO_2 in the atmosphere was disturbed. That lowered the Earth's albedo. Then, permafrost was no longer "*perma*nent;" it began to melt.

And here begins another, possibly more serious, feedback loop. As permafrost melts, the organic matter of which it is composed (dead plant and animal material) begins to decompose. This decomposition releases huge quantities of CO_2 and other far more potent greenhouse gases, like methane. Methane concentration in the atmosphere has grown by about 12% since 1974, from 1.65 parts per million (ppm) to about 1.85 ppm. Even though this is a slower rate of growth than for CO_2, Methane (CH_4) is a far more potent greenhouse gas; each molecule of CH_4 keeps more heat from leaving the earth than does CO_2 and the earth's blanket just keeps getting warmer. Melting permafrost releases CO_2 and CH_4 right where they can do the greatest damage in terms of feeding the feedback loop. The increases in CO_2 and CH_4 right above the permafrost mean that more sunlight will be absorbed, the permafrost will get warmer, the air will get warmer, more permafrost will melt, more greenhouse gases will be released, more snow and ice will melt, and so on. This runaway feedback loop has become very serious, very fast.

LOSS OF ARCTIC ICE

It is no surprise, then, that it is in the Arctic that we have seen the most dramatic effects of global warming. Arctic sea ice cover is declining at the rate of 12.5% per decade.

Up to 1980, the exposed water in the Arctic in September was the area in black in Figure 3. By 2000, the ice cover had shrunk to the dashed line and by 2012 there was little ice cover by the end of the summer.

Figure 3. Arctic ice cover over time
The area of Arctic ice diminished to the lowest level seen since satellite measurements were begun, about 3.8 million square kilometers, in 2012.

CHANGES IN GLOBAL TEMPERATURE

It has become quite clear that the earth is warming up. Some countries, like Australia and the United States, have seen record high temperatures. While some areas have seen colder than usual winters, it is important to not confuse meteorology with climatology. Where colder winters have been reported, they are most likely the result of changes in ocean circulation patterns. One must be careful, therefore, to look at the entire planet, its oceans, land, and atmosphere, taking the time to eliminate false readings that may reflect "urban heat islands" (the result of local changes in the built environment that produce high readings not representative of changes in the broader environment) to present an accurate portrayal of the trends in global temperatures. Climatologists can now assemble data from outer space and ocean depths, and all points in between, to present their best composite of how the world's temperature is changing. They can also extrapolate these data and add other sources of information such as tree ring data and ice core data to get a better picture of how the climate has changed over time. The chart below shows an aggregation of data reporting on likely temperatures over the last 2,000 years. The top and bottom lines are the high and low ends of the various data sets, and the middle line approximates what the most likely temperature was at a given time.

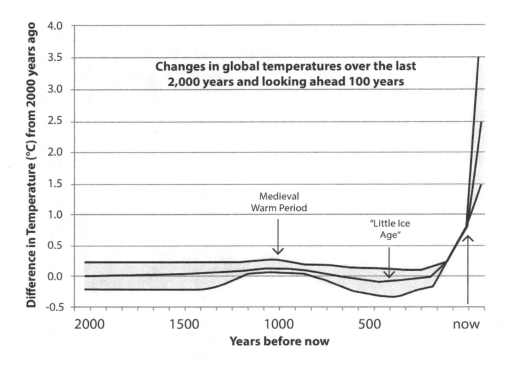

Figure 4. Changes in global temperatures over the years

You can see that we have quite a good idea of what the temperature is in our era and the uncertainty is large in both the past and the future.

Despite the uncertainties in the data, a few things are clear: the Medieval Warm Period and the "Little Ice Age" took hundreds of years to develop and pass, and the variations were surprisingly small. When we look at today's temperatures, we see that they are already well above those of the Medieval warm period and rising quickly. With both the Medieval warm period and the "Little Ice Age," many species had the time to adapt. Present temperature changes are occurring much faster than earlier ones and, although some species are trying to adjust by moving away from hotter areas towards either the poles or higher elevations, many will become extinct because there will be insufficient time to adapt.

REDUCING AIR POLLUTION

There are solutions to atmospheric pollutant problems, but none are easy to implement. We can, however, deal with local problems locally; issues like smog and particulates that do not become widely dispersed. The World Health Organization states that 2.4 million people die each year from causes directly attributable to air pollution, with 1.5 million of these deaths caused by indoor air pollution. This seems a rather conservative estimate when considered in context of another United Nations report, that of the United Nations Environment Program (UNEP), which concludes that 2.4 million lives could be saved each year just by controlling soot.

While some jurisdictions have rigorous measures in place to remove pollutants from industrial and transportation sources, others do not. Until all jurisdictions adopt such measures there will continue to be unnecessary suffering and death. Those who should take a leadership role in this area, and do not, are guilty of deliberately wasting the health and lives of those they should be caring for. You may recall the earlier comment about the steel company president who said that one can't make steel without making a mess. The steel industry will continue to profit from its deliberate contamination of the air if local leaders refuse to act.

Part of the difficulty in dealing with air pollution is that local leaders fear that taking action locally will put them at a disadvantage globally—and they are right. Today's response is to allow people to suffer and die prematurely because if steps are taken to protect the atmosphere and lives in one part of the world someone will open another plant elsewhere to kill people in that vicinity, and in so doing, take jobs away from the first community.

Clearly, the optimal response would be to have global pollution standards for all industries and transportation systems so that all would be on an equal footing.

The broader greenhouse gas problem presents an opportunity for global solutions. One might begin by recognizing certain rights and responsibilities, and then combine market forces with a cost for bad behaviour, in terms of atmospheric carbon pollution, and a reward for good behaviour. In such an approach, one begins with the recognition that the problem is a global problem and that everyone has an equal right to air unsullied by others. Next, the global community needs to agree on what is a tolerable level of greenhouse gases in the atmosphere and compute how much each person might emit, on average, before that level is breached. Then one could ascertain the *per capita* emissions in each sovereign jurisdiction. People in jurisdictions behaving badly in terms of *per capita* carbon pollution would have to contribute to a global "low-carbon energy development" pool of funds. The funds would be used primarily to develop and implement solutions for economic development and clean energy production in those jurisdictions where per capita carbon emissions are low. Beneficiaries of the fund would receive assistance in proportion to the extent to which their *per capita* emissions are below the global average.

Of course, one can anticipate the screams of outrage at such a proposal. People may be prepared to pay a tipping fee at the local dump, to keep solid waste off residential and agricultural lands. One may also be prepared to pay for sewage treatment and industrial wastewater treatment so as to not foul drinking water, recreational resources, and fisheries habitat. But people refuse to pay for waste disposal by emissions into the atmosphere, even though bad behaviour in this area may well be the most damaging of human waste disposal activities.

To those who are opposed to such solutions because they constitute a redistribution of wealth, to which they are philosophically opposed, I would suggest they review their insurance policies. In Canada, the amount of insured damage resulting from extreme weather increased by a factor of six between 2006 and 2012. Insurance premiums rose in tandem. This is clearly a form of redistribution of wealth after the fact: far better to take the necessary steps to prevent the disasters that loom than to pay for repairs afterwards, to say nothing of the reduction in human injury and death.

Proper atmospheric disposal pricing would soon lead to changes in behaviour at all levels of society. Those who argue that the costs associated with keeping the atmosphere clean would devastate the economy need to think more deeply about the economy. Although this will be dealt with more fully in later chapters; Chapter 22, "National Solutions," and Chapter 23 "International Solutions," suffice it to say here that the economy will survive and thrive. It is just that money will be spent in different ways. Instead of making and selling more and bigger things, people will be encouraged to make *better* things.

The effectiveness of the "economics approach" to protecting the environment is illustrated in those jurisdictions where individual households must contract for their own solid waste disposal. Where they might otherwise have dumped everything, they now have an economic incentive to reduce their waste volume by composting their kitchen scraps. This economic incentive to do the right thing also protects the environment.

PUTTING A PRICE ON ATMOSPHERIC DISPOSAL

How can we approach the ideal of putting a price on waste disposal into the atmosphere?

First, every jurisdiction must recognize the need for action. This is the case whether a given country is responsible for two percent of the problem or twenty percent. If we cannot accept our responsibility to save the atmosphere in its own right, perhaps we can act to reduce the millions of lives deliberately sacrificed to air pollution each year and the huge sums of money spent on related health problems. Countries or regions that evade the issue on the basis that they are but a small part of a big problem are guilty of a dangerous cop-out. To illustrate, let me paint a word picture in the nature of a tipping-point analogy.

Let's say you are walking along the edge of a beautiful canyon, and you come to a plank stretching out over the canyon but firmly planted on the cliff edge by means of a large boulder. To get a better view of the canyon, you venture out on the plank until you feel the board begin to wobble. Your weight at the end of the plank is beginning to overtake the counterweight of the boulder, and the entire assembly is about to tip into the canyon, taking you to your death. What do you do? You turn around and walk back, of course.

Now imagine that you are hiking with a group of fifty hikers and you come upon an immense plank secured by many boulders on the cliff's edge. You all decide to step out to enjoy the view. At some considerable distance from the edge you feel the plank beginning to move. What do you do? If you alone turn back, the problem will not be solved because you will not get to safety before the entire assembly slides over the edge. As you are only 2% of the problem, your individual action won't prevent the disaster. You might try to push others off the plank to increase your chance of survival, but that is hardly an ethical solution to the problem. In the real world, the globally warming world, that means that the rich world would watch dispassionately as drought or more frequent and powerful storms ravage those less fortunate—the inevitable result of an unwillingness to make the social, philosophical, and economic changes needed to stop the deliberate waste of our atmospheric environment and to begin to restore it to its former health. Eventually, such selfish behaviour will affect everyone. But then it will be too late to salvage the situation.

In our tipping-point analogy over the canyon, the alternative is to persuade the other 49 hikers on the plank to cease their reckless action and turn back before it is too late. It may not be an easy task, but there is no other ethical resolution of the problem. Safely off the plank, you had better put up a sign warning those who follow not to venture out on such a destructive path again.

One of the reasons the proper course of action is so difficult is that there are powerful interests acting against it. Those who mine ancient carbon resources and produce cheap energy from those resources have a lot at stake and will fight threats to their profitability at every turn. Their counterargument will begin with the denial that global warming is happening, and even if it is, it's not because of human activity.

They would prefer to blame something else, like the sun; hence the quote with which I began this chapter.

The best way to counter such denial, and avoid falling into the atmospheric abyss, is to inform yourself and ask such challengers a few straightforward questions:

> Has the Earth changed its orbit? No. While this may account for some earlier climate changes, we cannot attribute the current one to that cause.
>
> Has the sun's energy increased over the last couple of centuries? No. In fact, the sun's energy output seems to have dropped very slightly over the past few decades while the planet has been warming up.
>
> Has the tilt of the Earth shifted? No. Some scientists think this could account for earlier periods of planetary warming, but not the one we are now experiencing.
>
> Is there any variable to which the current atmospheric warming can be related? Yes. Global temperature changes correspond roughly to human carbon emissions.

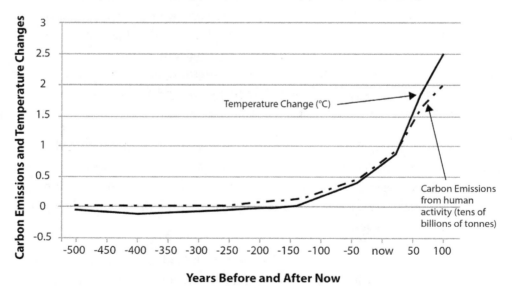

Figure 5. Carbon emissions and global temperatures

So, to paraphrase the mythical Sherlock Holmes, with his impeccable deductive logic, "…when you have eliminated the alternative possible causes, the one that remains must be the truth." Despite this, the anti-environment-industrial-complex will exclaim: "But there are some scientists who say it isn't so!" To respond to this argument, consider the situation where you have been diagnosed with stomach cancer. Nine out of ten gastroenterologists look at your charts and conclude that you have a tumour the size of a golf ball in your stomach and recommend surgery. The tenth suggests that you might have swallowed a golf ball. Of course, you can take your chances with the diagnosis of the tenth—but would you?

From a political/economic point of view, Canadians might find the effects of global warming beneficial. The Arctic ice cap melts, so they can open the Northwest Passage through the Arctic to transportation. They may drill for oil in the Beaufort Sea, and profit as they add even more carbon to the atmosphere. The permafrost will melt. The boreal forests will move north, with the grain belt and corn belt moving in behind. That should be profitable! Just dig up those tar sands and get the heavy crude to market as fast as you can. Let the present crop of politicians and oil companies benefit. Pass the environmental bill on to future generations.

Russia and Greenland could also benefit significantly from this deliberate atmospheric insult. Meanwhile, Australia can burn—northern Africa, too. And southeast Asia, perhaps parts of Europe as well, like Spain and Greece. Those who profit by the damage are not those who suffer it.

While Europe is in the forefront in trying to come to grips with carbon reduction, others are not. Consider the words of the American president when addressing the first international attempt to deal with the problem. On June 11, 2001, President George W. Bush presented America's position with respect to global climate change. Among other things, he said:

> "Our country, the United States, is the world's largest emitter of manmade greenhouse gases. We account for almost twenty percent of the world's man-made greenhouse emissions […] We recognize the responsibility to reduce our emissions […] Kyoto is, in many ways, unrealistic. Many countries cannot meet their Kyoto targets. The targets themselves were arbitrary and not based upon science. For America, *complying with those mandates would have a negative economic impact*, with layoffs of workers and price increases for consumers." (emphasis added)

In the intervening years, the United States has continued its substantial dumping of carbon wastes into the global atmosphere.

Our atmosphere is the most crucial and the most shared component of Earth's environment. We continue to deliberately contaminate the atmosphere with greenhouse gases, particulates, toxins, and unnecessary stray light and sound pollution. We do not manage this resource in a rational way, where those who cause the damage are held accountable, and those who are most at risk because of the damage are compensated.

The response of the carbon industry, and those jurisdictions that rely most on it for economic growth, has been to attack the science and those who accept it. Further, they have mounted a counter campaign, focussing on the evil of a Carbon Tax. It is no more a carbon *Tax* than disposing of your rubbish at the local dump is a rubbish tax. You pay a fee at the local dump to dispose of your rubbish. The dump operator, hopefully, finds a way to handle that rubbish to the minimum disruption of the environment. So, let's replace CARBON TAX with what it really is—a CARBON DISPOSAL FEE. In particular, if you want to burn Carbon and pollute the atmosphere with your waste, you should pay a fee that can be used by the rest of society to clean up your mess.

Next, let's reward people for burning less carbon. The Government charges the disposal fee at the gas pump, say, and gives each citizen the same portion of all the disposal fees collected. If people want to save money, they burn less carbon by driving a smaller car and driving less often, so they pay less, but still collect the return.

THE COST OF CONTINUING TO POLLUTE

We have become addicted to using hydrocarbons as an easy way to get energy. Now some people are suggesting that we "geoengineer" the world so that we can continue the addiction while we do something extra to try to counteract it. That is like dealing with a drug addiction problem by prescribing the life-long administration of an antidote. You don't get the addict off the drug, but you hope someone will be able to pay for the antidote forever. Geo-engineering solutions proposed by some people are like trying to put our planet on life-support before we have tried to cure, let alone prevent, the hydrocarbon addiction disease from which we suffer. Like life-support in the medical analogy, this will be much costlier, and unsustainable in the long-term. But then, the present generation can defer those costs to subsequent generations. As long as we feel we can avoid paying for our misbehaviour we will not take the proper action at the proper time.

The concept of a "500-year" storm was developed to describe the kind of storm that would strike a given area only once in 500 years. Hurricane Harvey, which struck Texas in 2017, was the *third* such storm to strike Houston in the last *40* years. The number of weather-related disasters has increased from about 200 a year in the 1980s to more than 600 a year in 2016, according to one of the world's largest reinsurers. So, we have chosen to pay the cost of treating the illness rather than preventing it, even though the prevention option is likely to be the less costly alternative.

About 90% of the additional energy added to the atmosphere winds up in the world's oceans. Hurricanes and typhoons draw their energy from the heat of the oceans. The heat content of the world's oceans is the temperature of a cubic meter of ocean water multiplied by the number of cubic meters of sea water at that temperature. With about 1.37 billion cubic kilometers of sea water (yes, that is 1,370,000,000,000,000,000 cubic meters) in the oceans, it doesn't take much of a temperature rise to increase ocean heat content dramatically. Warmer oceans produce more intense and longer-lasting storms. This, in combination with higher sea levels from melting glaciers and expanding oceans (water expands when heated), produces storm surges that move farther inland. These problems will get worse and worse over time.

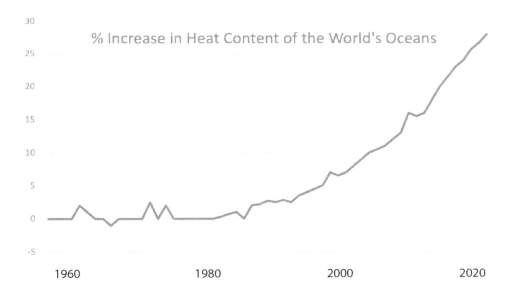

Figure 6. Increase in the heat content of the world's oceans

Imagine a couple in their thirties about to start a family. In thirty years (2050) their children may want to have children. When those grandchildren have grown up and are thinking of starting a family (2110), they will be coping with the damage caused by global warming, with trillions of dollars being spent to cope with horrible storms, droughts, wildfires, coastal erosion, and inhabitability along the coasts. It would not be unreasonable for them to look back to the 2020 era and ask: "How could our grandparents have been so selfish? They knew that they were moving billions of tons of carbon, that had been stored in the ground for millennia, into the air in just a few decades. They knew what would happen as a result. They continued to drive big gas guzzlers and fly around the world. Now *we* must pay. *What were they thinking?*"

Still, some of the countries that caused the bulk of the problem over the last two hundred years refuse to accept their responsibility to pay for the damage they have caused, calling it an unfair burden, even though they continue to enjoy the benefits of their past bad behaviour.

There was a meeting of heads of governments in Copenhagen, Denmark in late 2009 to advance the objectives of the Kyoto Protocol on climate change. It was a disaster. A subsequent meeting under the auspices of the United Nations in Cancun, Mexico, in 2010, made some headway and annexed the Copenhagen results to its own. It is generally acknowledged that pledges made in the Copenhagen accord and given enhanced status in Cancun are nowhere near good enough to limit global warming to two degrees or less. These objectives will never be met if countries such as China and America refuse to accept legally binding targets. Many observers believe that if one seeks hard targets from such countries, the whole UN process will fall apart. Once again, we are setting out to deliberately fail our environment.

One reason that some countries, such as Canada, do not take their CO_2 reduction responsibilities seriously, is that their politicians think their country will enjoy more economic benefits than costs, and they don't care about the broader global effects of bad CO_2 behaviour. A 2010 publication "Degrees of Change: Climate Warming and the Stakes for Canada" by the Royal Canadian Geographic Society and the National Round Table on the Environment and Economy (NRTEE) urges Canadians to start thinking in terms of "climate prosperity" along with climate change. After all, northern temperate countries are likely to experience less harsh winters (and lower heating bills) along with longer growing seasons. David McLaughlin, president of the NRTEE, is quoted as saying, "There are risks to climate change, but there are also opportunities." He accepts that climate change "is happening and we can deal with it, and we can gain from it." The "we" in this case are Canadians. The pain suffered by a broader "we," residents of the rest of the world, is apparently not of concern.

"Degrees of Change: Climate Warming and the Stakes for Canada" delivers a colourful graphic presentation showing the risks and benefits to Canada. With a two-degree increase, we will see reduced prairie grain crops because of desertification, but there will be some offsetting increasing crop yields elsewhere. The health of the Great Lakes and the fish therein will suffer, but there will be more codfish north of the 60th

parallel. The ski season in southern Canada will decrease by 15 to 25%, but the golf season will expand by 7 to 20%. Runoff in the South Saskatchewan River basin will decline but there will be a shorter winter road season.

A three degree increase in global temperature would wipe out the Canadian Atlantic fishery but provide easier access to northern oil and gas (so it will be cheaper to produce more greenhouse gases) according to "*Degrees of Change.*" It would also double the frequency of Prairie droughts but increase runoff in northern and eastern Canada. Western glaciers will diminish by 50%, reducing runoff, but there will be twice as many extreme rain events, with the associated flooding costs, across the country. At four degrees, the Halifax waterfront will disappear, along with 20% of the boreal forest and 10% of the tundra, and Canadians can expect two to four times as many forest fires. (These last two predictions, if true, will add to the runaway greenhouse effect.) The four-degree increase will also make the Northwest Passage more navigable.

Just before the Copenhagen meeting, the computers at a leading climate change institute were hacked into and a very selective group of emails were extracted and published, purporting to show that there was a cover-up of data which ran counter to the prevailing scientific view on the subject. If the intent of those who hacked into the computers was to discredit climate change science at the outset of the Copenhagen conference, their attempt was very successful, despite the statements of scientist at the conference that the leaked documents changed nothing. Independent reviews were launched on the question of whether the leaked documents affected the conclusions of the International Panel on Climate Change (IPCC). These reviews found no evidence that weakens the case for actions that would limit carbon emissions, but by the time the reviews were published, the damage had been done.

Part of the problem, as I learned in my days as an environmental law judge with the Ontario Environmental Assessment Board hearing sworn evidence from world experts on a wide-ranging variety of environmental topics, is that the scientific method requires researchers to make clear the uncertainties of their work. Science skeptics feel no such compunction. Thus, if Dr. X accepts that a 19 times out of 20 statistical reliability test is necessary to conclude that a temperature trend is likely to hold true, and his research finds that the reliability is only 18 times out of 20, he is bound to conclude that he cannot say, with scientific certainty, that the trend will hold true. His opponents seize on this admission about statistical certainty as evidence that there is no trend, which is quite another matter. Non-science people may be persuaded to accept the opponents' statements that there is no warming trend because of a statistical failure to meet a most stringent reliability test.

Another part of the problem is the position of many in the media who are obliged to present both sides of an issue without qualifying whether one side has a thousand times the support of the other. This leads to the kind of reporting where, during a cold snap in central Europe, say, a newspaper will report that: "Contrary to the report of the IPCC, Dr. X, an esteemed meteorologist at our local university, says that there have been global warming periods in the past and this is no different." This puts Dr. X and the IPCC on equal footing, even though Dr. X is a meteorologist, not a climatologist, and there are thousands of qualified scientists who would challenge Dr. X's unscientific finding in support of the findings of the IPCC.

The primary conclusion of the individual scientist in this case, that the current situation is no different than historic climate changes, is wrong in many aspects, not the least of which is that mankind has not had the power to materially affect the concentration of gases in the environment until the modern era.

Because of my scientific training and former work in the field of adjudication of environmental matters, I often am asked by friends, acquaintances, and others to respond to the attacks on the work of the IPCC. I generally encourage them to dig deeper, to check the qualifications of those making the statements and the balance in the reporting. If you have chest pains and your cardiologist shows you the results of a battery of tests confirming that you need bypass surgery, you will not be helped by a podiatrist recommending that insoles in your shoes will cure the problem (especially if the podiatrist profits from the sale of insoles).

For an independent corroboration of the pace of global warming one only need check the pace of migration of various species into areas that had previously been too cold for them. That often means migrating to higher altitudes. In the northern hemisphere, it also means moving north, and in the southern hemisphere, moving south to escape tropical heat. Chris Thomas of the University of York has recently published a comprehensive scientific paper on species migration driven by global warming. Having examined about 2,000 species he finds that they are moving away from the equator at more than one and a half kilometres a year, on average. Various species are also moving up the slopes of mountains at a rate of more than a metre in elevation each year to find new warm homes at higher altitudes or to escape the heat at lower altitudes.

Those opposed to the findings of the IPCC are powerful and well-versed in the use of public relations to transform public views regardless of the underlying science. As James Hoggan, a public relations consultant, puts it: "For years, there have been these kinds of campaigns that are aimed at manipulating public opinion […] It happened with the tobacco industry. It happened with the chemical industry. It happened with the asbestos industry. And now it is happening with climate change."

The last great global warming occurred about fifty-six million years ago. The PETM (Paleocene-Eocene Thermal Maximum) resulted from the release of carbon as the old supercontinent Pangea split apart to form the continents we see today. Geoscientists believe this process occurred over a period of about 20,000 years, and it took nearly 200,000 years for the earth's natural buffers to bring the temperature back down to a normal level. This is an example of one equilibrium being replaced by another because of geological processes, and then finding yet another equilibrium as the earth cooled. Even though that warming was only 5°C and was spread out over 20,000 years many species became extinct. At the rate we are now heating the planet, we could well warm the planet by 5°C in a hundred years, and many more species will be unable to cope with such a rapid change.

As unsettling as the projections of global warming this century are, they are only half as bad as they might be if the projections of Massachusetts Institute of Technology (MIT) scientists hold true. They factored in economic growth in China and India and updated data on the rate at which the world's oceans take up both heat and carbon dioxide and concluded that global land and ocean surface temperatures would likely rise by 5.2°C, more than twice the 2.4°C projected by earlier models. Most scientists believe that an increase of just 2°C would have quite serious implications for the survival of many on planet earth.

GEOENGINEERING

To mitigate the rise in temperature, some turn to geoengineering. Geoengineering is the term used to describe proposals to reverse the harmful effects of human activity on a global scale by additional deliberate human activities. It is most often used with respect to the anticipated negative effects of global warming, where it is hoped that we could fix our global warming problems after the greenhouse gases that caused it have been emitted, rather than reducing or preventing those emissions in the first place. This would be another deliberate step to change our environment whether or not we fully understand the implications of doing so.

Several such proposals are already on the table. Shooting sulphur particles into the air, or dispersing sulphur dioxide or sulphuric acid at high altitudes is a favoured approach. The idea here is that if enough sulphurous particles are injected high in the stratosphere, they will reflect some of the sunlight back into space before it gets into the lower atmosphere or strikes the earth to warm things up. Under these types of proposals, we can go on emitting greenhouse gases and just keep adding more and more sulphur to compensate. The sky will become hazy, the stars will no longer be clearly seen, but we could get some spectacular sunsets.

The methods proposed for geoengineering the atmosphere sound like science fiction today but could become reality if humans decide to continue to deliberately abuse the environment. One proposal is to use large artillery to shoot sulphurous particles directly into the atmosphere. Another is to construct giant chutes held aloft by helium balloons, and pump sulphurous compounds directly up these chutes. Given the resistance to proposals like windmill farms that benefit the environment, but have a negative visual impact, one can only imagine the difficulty of finding a site for a "chute farm" that reaches up to the stratosphere. More plausible proposals include using a fleet of jumbo jets or other heavy lift aircraft to continually ferry such materials as high as possible. It has even been proposed that huge hybrid plane-blimps be used to lift heavy loads of sulphurous materials into the stratosphere, where the rarefied atmosphere doesn't support normal aircraft.

These solutions are like trying to deal with the equivalent pile of manure described at page 16, and earlier, by offering your neighbour a gas mask with rose-coloured lenses so that he couldn't see or smell the problem, rather than doing something to prevent it.

Unsurprisingly, proposals to deal with pollution by deliberately adding more pollutants could have serious negative environmental effects. They will certainly have unintended consequences. But some of the consequences are already known or at least predicted by current modelling techniques. One of the expected consequences would be the continued destruction of the ozone layer—the invisible high-altitude screen that protects terrestrial life from ultraviolet radiation. Another anticipated consequence is a reduction of rainfall in certain areas where it is desperately needed. Mathematical models suggest that rainfall reductions in the southern food belt of China could be severe; this in a country that is already looking abroad for alternative food supplies for its huge population.

Geoengineering is attractive because of economics. Spending a couple of billion dollars a year on sulphur aerosols to offset warming rather than spending hundreds of billions, or a few trillion, dollars to deal with the problem by reducing carbon inputs at source forever, is the easier solution for the present generation. If we go that route, it will be another example of deliberate actions that waste this planet's environment.

There are many downsides to geoengineering approaches, not the least of which is what happens when international support falters after a few decades of unrestricted carbon emissions. This could arise because of economic factors or because it is discovered too late that there are unintended consequences for the ozone layer or rainfall in specific agricultural areas around the world. If China were to be a major contributor to aerosol injection and there was a related drop in food production in India, the subsequent tensions are obvious to predict.

Other approaches, such as fertilizing the oceans to accelerate photosynthesis in phytoplankton, are not only unproven, but they might also have a whole different set of unintended consequences as well.

I used a metaphor earlier to explain why we need to try to persuade everybody to respond properly to the global warming danger before it is too late. That was the "plank over the chasm" metaphor, where failure to retreat from the end of the plank would result in disaster. In that metaphor, sceptics said, "The rock's not really letting go—it's just a tremor. Don't worry. Let's keep going." Others said, "You know, the records show that the rock trembled many times before and there weren't even people around to make it tremble. It's not us. Let's keep going." In today's warming world today, sceptics say, "It's not really a warming trend—it's just a blip." Others say, "You know, the records show that there were warm periods many times before and there weren't even people around to make it warmer. It's not us. Let's keep burning carbon."

The problem with the metaphor is the time scale. When the board is going to let go, you had better persuade your fellows to turn around fast or all is lost. With global warming, the time scale is hundreds of years. We'll survive. Our kids will probably survive, and we hope our great, great grandchildren will figure out how to live with the mess we are making. In the interest of future generations of all species on earth, we must persuade those walking this global warming plank with us to turn around and find a lifestyle that puts less of these warming gases into the atmosphere. This doesn't mean going back to a hunter-gatherer society, but rather finding an alternative lifestyle that relies less on burning carbon.

Recall the words of US President George W. Bush in the summer of 2001:

> "Our country, the United States is the world's largest emitter of manmade greenhouse gases. We account for almost twenty percent of the world's man-made greenhouse emissions. We also account for about one-quarter of the world's economic output. We recognize the responsibility to reduce our emissions […] This is a challenge that requires a hundred percent effort; ours, and the rest of the world's."

In the intervening years, the United States has continued to emit between six and seven billion tons of greenhouse gases as CO_2 equivalent every year. That's about sixty-five billion tons since the statement was made. A hundred percent effort?

Just because we *can* dump millions of tons of pollutants into the air each day doesn't mean we *should*.

Now let's turn our attention to the next most critical component of our environment, water.

Water

"You really fucked-up up this time, Kingham," said the Senior Assistant Deputy Minister (SADM) as he raked me over the coals in his plush Ottawa-area office. The obscenity of the SADM's language was matched by the obscene message he was transmitting. The essence of his message was that keeping drinking water safe was not important if doing so could jeopardize economic growth. In my dual roles of Chairman of the Great Lakes Water Quality Board and Regional Director-General for Environment Canada in Ontario, that message ran counter to everything I had been working for over the preceding decades.

I choose not to speak obscenities, and apologise for writing the opening sentence above, but it is a direct quote. I had been summoned to get on a flight from Toronto and go to the SADM's office immediately, with neither adequate notice nor any indication about the purpose of this "urgent" meeting. A year earlier, the Canadian Minister of the Environment and I had met with our counterparts from the USA, in New York City. The Administrator of the United States Environmental Protection Agency (the EPA) and his Region 2 Regional Director represented the US side of the Niagara River; the Canadian Minister of the Environment and I represented the Canadian side. The Minister and the EPA Administrator instructed us to come up with an action plan to clean up the toxic chemicals lying alongside the Niagara River and leaching into that river, poisoning the drinking water of millions, mostly Canadians, downstream. Almost all of the toxins were coming from dumpsites on the American side of the river. Love Canal and Hyde Park, overlooking the Niagara Gorge, where some 80,000 tonnes of dioxins and other chemical poisons were buried and leaching into the river, were the most egregious sites. We were to find a solution for that part of our wounded environment.

Recognizing the importance of this area to the water quality for people from Toronto to Montreal and everywhere in between, I had proposed to the Minister of the Environment that we should create a Niagara River Toxics Action Plan. The outline of that plan was the stimulus for the Canada/US meeting. A year later, our side had a plan for the relatively minor pollution coming from Canada. Then I got a call from a reporter with the Buffalo Evening News asking what I thought of the American Plan.

While the Canadian side had developed step-by-step measures to prevent toxins from getting into the Great Lakes, the US had just stapled together their existing programs from the preceding decade and called it their "plan." I called it "meaningless."

The next day the EPA Administrator met with Canadian Ambassador. The EPA was furious about my criticism of the US "Plan." The Canadian Ambassador called the Canadian Prime Minister, who called the Environment Minister, Suzanne Blais-Grenier. The Environment Minister called the SADM, and there I was, being sworn at in his office. The SADM made it abundantly clear that the only thing important to the Canadian Government was the NAFTA free-trade agreement, and that the government didn't care about toxic chemicals in the Great Lakes.

I have devoted some space to this event because it crystallizes the dilemma of environmental damage in the context of economic growth: a dilemma that I will revisit at some length in later chapters. Whenever businessmen and the politicians they control perceive that steps to protect or preserve the environment will interfere with their plans for economic growth, they will vigorously condemn such steps as being against the interests of their customers or constituents. This conflict of views is as obvious in the water environment as it was in the air environment. Drawing attention to a deficiency in environmental protection when there was a trade agreement on the table was "just not on." Risking human health because of toxins in the drinking water was okay if there was a chance for enhanced trade.

Speaking truth to power led to my demotion and eventual departure from Environment Canada, after twenty years of hard work trying to save the environment from within the system, rather than attacking it from without. It was not the first time a scientist was sacked for speaking out on environmental issues in order to serve the public; sadly, it will not be the last.

Unlike the air environment, where the concern was almost entirely about the quality of that component of the global environment, the water environment offers many examples of deliberate waste of the quantity of water available as well as damage to its quality.

WATER QUANTITY ISSUES

Planet Earth is sometimes referred to as "the blue planet" or "the watery planet" because there is more liquid water on our planet than on any other known planet. Despite its abundance and importance to our survival, we do not know precisely where it came from. A small portion may have been tightly bound up in the minerals of which our early, hot planet was formed. Any of that water released in the early days, a few billion years ago, would have boiled away. As the planet cooled, more recent releases from such minerals would have been able to stay around as water vapour and liquids. But by far the largest portion of water on Earth seems to have been delivered by comets and asteroids. Since then, it has been recycled over and over again, thanks to the wonder of the hydrologic cycle. The exact origin of our water supply is less important than what we have done with it, and to it, in the last few centuries.

It is not possible to state a precise figure for the total amount of water on Earth, but 1.4 billion cubic kilometers is a pretty good estimate. That's a very large number, and it is quite difficult to appreciate just how big it is. Trying to understand it in terms of so many Olympic-sized swimming pools or football fields filled to a certain depth won't help much either, as the number will only get bigger.

A box with equal sides that would hold 1.4 billion cubic kilometers would have each edge as long as the distance between Toronto and Minneapolis, or between Paris and Rome. So, imagine drawing a line on a map between either of those pairs of cities and constructing a three-dimensional box using that line as one edge.

Although that seems like a lot of water, 90% of it is salt water in the oceans, and it's not easily available for drinking and agriculture. That leaves us about 140 million cubic kilometers of fresh water. Our box shrinks to one having an edge about half as long as the first box. When we then remove water that is stored as ice, the box shrinks to one just 241 km on edge, less than the distance from Paris to Brussels or Toronto to Kingston. We can't rely on all that ice melting to give us fresh water because most of it will run into the ocean and become saline.

The problem is, a lot of the remaining fresh liquid water is in the atmosphere, in plants or in very deep aquifers. What's left over is about 420,000 cubic kilometers of readily available fresh water—a box about 75 km on each side.

There are about eight billion people on this planet. If each person gets an equal share of the readily available fresh water, he would have about 50,000 cubic metres of water (13 million gallons); that is, a box of water with edges about 40 meters long (130 feet) sitting on about one eighth of a hectare (one third of an acre) of land.

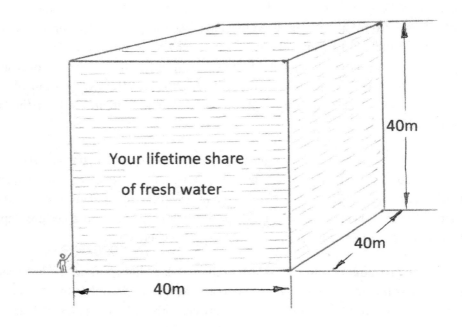

Figure 7. Each person's share of available fresh water

For a perspective you can hold in your hand, take a one litre measuring cup and fill it with water. That's 1,000 millilitres. Imagine that is all the water in the world's oceans, glaciers, aquifers, lakes, rivers, and atmosphere. Now, pour 900 millilitres into the sink. That's the portion of water in the oceans. (You could do this example in ounces, but it is so much easier in the metric system.) You have just 100 millilitres of fresh water left.

About two-thirds of that fresh water is in the ice caps of Greenland and at the north and south poles, and in glaciers on mountain tops. So, pour out another 67 millilitres until you have only 33 millilitres left. That's the amount of fresh water in lakes, rivers, plants, aquifers, and the atmosphere.

Now, pour out all the remaining water from the one litre measuring cup. There are probably three or four drops of water still clinging to the walls and bottom of the cup. That's the portion of the Earth's fresh water available to our eight billion people in lakes, rivers, and groundwater after you remove the "plants, deep aquifers, and atmosphere" part. That amounts to about three twenty-thousandths ($3/20,000^{ths}$) of the total.

Instead of thinking of a beautiful blue globe suspended in space, or a large box of water, when you think about the portion of water available to all humans on Earth, think instead of that empty litre jug with just a film of water remaining, and be mindful of the great care we must take of that very limited and precious common resource.

For now, returning to your personal box of water, let's consider what you might do with it. For example, you might decide to water a typical North American golf course for three quarters of one season. That's it. All gone. No more water for the rest of your life.

The average total water consumption per person in Canada is about 1,130 cubic meters per year (m^3/y). In the United States it is about 1,630 m^3/y whereas in Denmark it is only about 180 m^3/y. If everyone had the same amount of water to start with, the typical Canadian would run out of water in about 53 years and the typical American in 37 years, but a Dane would still have water left over after 300 years. Most of the rest of the developed world is somewhere in between. For most of the still developing, or emerging, economies the average *per capita* water consumption is well below that of the developed world: Asians use about 500 m^3/y; South Americans about 300 m^3/y; and Africans less than 200 m^3/y. When one adds the water which is used to make products consumed in different countries the situation appears more dire. The so-called "water footprint" is more than 3,000 cubic meters per person per year in the US, and less than 1,000 cubic meters in many developing countries. That means that the average American would run out of water before age 20 if he were restricted to only his share (50,000 cubic meters) of the world's fresh water.

While our consumption of readily available freshwater often exceeds supply, the supply is dwindling. Part of the reason is that we are pumping water out of deep-water aquifers at a faster rate than it can be replenished, and global warming is reducing water supply to many rivers in the world. As the planet warms and glaciers melt there will be even less glacier melt water to slake our thirst, water our fields, and feed our industries. When the glaciers have melted completely, the streams and rivers they feed will dry up. Imagine the disastrous effects on countries like India and Pakistan where people rely on the continuous supply of meltwater from Himalayan glaciers for survival.

Elsewhere in the world, losses of water reserves are huge. Between 2003 and 2010, parts of Turkey, Syria, Iraq, and Iran, along the Tigris and Euphrates rivers, appear to have lost 144 billion cubic metres of stored water. While some of this was due to evaporation and the dehydration of soils, most of the loss, about 90 billion cubic metres, resulted from depletion of groundwater reserves. The government of Iraq alone drilled about 1,000 new wells into its aquifers to weather the drought of 2007.

Even in North America, the threat of insufficient water looms. Consider the Colorado River, for example. Yearly flows in that river range from about 6 billion cubic metres to about 25 billion, with a long-term average of about 18 billion cubic metres. The sum of water allocated to various users plus evaporation is about 21 billion cubic metres a year, an excess of 3 billion cubic meters above the long-term average. Because some of those on the upper reaches of the river do not use the full amount allocated, the shortage is not as bad as it soon will be: the US Bureau of Reclamation predicts that by 2035, there will likely be a *shortage* of three billion cubic metres a year. Fifteen percent of the crops in the United States depend on the Colorado River for water, as do large cities such as Las Vegas, Los Angeles, Phoenix, Tucson, and San Diego, accounting for more than half of the water supply in many of these areas. What will they do when it runs out?

It wouldn't be so bad if all the water that is drawn was being used for essential purposes. The tragedy is that so much of it is being wasted by design.

How water is used depends on the economic status of the user. Worldwide, about 70% of human water consumption is for agriculture, 22% for industry and 8% for domestic uses. Use in high income countries is about 30% for agriculture, 59% for industry and 11% for domestic purposes. In low and middle-income countries, consumption patterns are quite different: 82% for agriculture, 10% for industry and 8% for domestic purposes. A great deal of water is wasted before it even reaches the user. Worldwide, between 25% and 50% of treated water is lost because of leaks in the water supply systems of cities.

The World Health Organization finds that water contaminated with sewage causes about 4 million cases of cholera a year, as a result of which about 100,000 people die each year. It is not "rocket science" to separate human sewage from drinking water and to provide safe, clean drinking water for everyone. But we have decided, collectively, globally, that it is more important to build space stations, travel to the moon, and explore the universe than it is to provide healthy drinking water here on earth. It is distressing that we can find trillions of dollars to wreak vengeance on those whom we believe have wronged us, but we choose not to find a fraction of that to improve the lot of the planet's citizens by providing clean water and eliminating cholera.

The largest use of water is for agriculture. Unfortunately, much of this use is unsustainable. Aquifers, the large natural underground storage vaults of water, are being drawn down faster than they can be replenished. River water supplies are being depleted. Water withdrawal for agricultural use exceeds supply in many places around the world. This is most pronounced in Northern India, Pakistan, and the southwestern United States. It also happens elsewhere, with China, Southern Africa, and Spain being some of the larger unsustainable consumers.

The depletion of aquifers is a classic tragedy of the commons. When a farmer pumps the limited hydrologic resource below his land without regard to the effect it will have on the availability of the resource to his neighbour, it is like draining a shared water canteen whilst on a long hike, without concern for the well-being of one's companion. The depletion of trans-boundary aquifers raises the insulting nature of this tragedy of the commons another notch. Worse still is when the person drawing the limited resource out of the ground uses it for a purpose well down the scale of those purposes essential for human survival. Thus, if the operator of a golf course takes aquifer water to keep his golf course green, and the crops on a distant but hydrologically connected farm wither as a result, the insult is obvious. When the distant but hydrologically connected "neighbour" is a resident in an adjacent country, there is a threat to peace as well as social order. Books have been written about the threat of war arising from environmental abuse of water supplies. Gwynne Dyer's book, *Climate Wars*, is a good example.

The previous chapter described how a warming atmosphere causes changes to the climate. Not surprisingly, climate change causes changes in the quantity of water available. At times, there is more water than can be handled, and areas are flooded for a short period of time. Too often, the warming climate results in too little water available for agriculture and other essential uses. Extensive research has shown direct links between climate change, the hydrologic cycle, and the demands humans place on water supplies. More than a hundred million Africans face water shortages. Agriculture in Africa is suffering severe negative effects because of water shortages arising from climate change.

Water is quite deliberately being wasted on frivolous things such as golf courses and holiday destinations developed in the middle of deserts. Much of the water used for agriculture is spray irrigation, where only a small fraction of the water drawn from limited supplies finds its way into food.

Just because we *can* squander the limited freshwater resources of the planet doesn't mean we *should*.

Jim Kingham

WATER QUALITY ISSUES

Sub-consciously, and consciously, water quality has been an intrinsically important part of my life since my earliest days. I grew up in a rural setting on the outskirts of a small Ontario town. Our water came from a well in a sandy soil layer thirty to forty feet beneath the back lawn. We knew to restrict our water use so as to not draw down the water level too quickly, lest it drop below the bottom of the well pipe. We also knew how critical it was that we protect the land above from any contaminant that might find its way into our well.

Next door was a dairy farm. With alternating crops of hay for winter forage and summer pasture for feeding his cows, farmer Hahn knew how to recycle the wastes from his dairy operation to fertilize his hay production without contaminating his pasture operations.

I enjoyed a childhood spent barefoot, up to my knees in the stream that ran through the woodlot at the bottom end of the Hahn property. Clear water flowed through that stream, nurturing a most diverse and beautiful collection of aquatic life. Summer holidays included swimming in the then sparkling waters of Lake Erie, or fishing for bass in the Grand River.

Not surprising, then, that my first professional job was to study water quality issues in the Great Lakes. Work to prevent water contamination from environmental emergencies followed, then work on the international convention to prevent dumping at sea, and work on conventions to prevent contamination from oil tankers and on marine environmental protection under the Law of the Sea Convention, and finally, work as the Canadian Chairman of the Great Lakes Water Quality Board.

I have learned that there is an awful lot of unnecessary water pollution in the modern world.

The simplest of human wastes and potential pollutants are unavoidable. We eat, we defecate, we produce potentially damaging wastes. The harmful effects of these wastes in our water supplies have been known for centuries, if not millennia, since the beginning of the time when humans began living near to each other as they grew away from their earlier hunter-gatherer lives. Although we have known that it is important to protect water supplies from sewage, we continue to allow such pollution.

Modern sewage pollution is not like it was a few centuries back. In the early days, one just had to be careful to not release too much untreated sewage upstream of water intakes. Sewage treatment plants were constructed to remove solids and reduce the chemical and biological oxygen-demand (COD and BOD) of sewage before discharging it into rivers, lakes, and oceans. The COD is the amount of oxygen consumed to chemically oxidize organic water contaminants to inorganic end products, and the BOD is a measure of the amount of oxygen required to remove waste organic matter from water in a process using oxygen-consuming bacteria. Sewage from livestock operations was used wisely for fertilizer to sustain operations.

Now, sewage discharges are laden with thousands of chemical components, some of which are toxic, some of which mimic hormones and disrupt reproduction systems in wildlife, some of which promote bacterial mutations which make bacteria resistant to antibiotics, and some of which feed algal growth in rivers and lakes. A clear illustration of this last effect occurred in the late 1960s when some of the Great Lakes, Lake Erie in particular, were being over-fed with phosphorus. Algal growth in the Great Lakes was phosphorus limited, meaning that although there was plenty of sunlight, nitrogen, and carbon around, the extent of algal growth each summer was constrained by the amount of phosphorus available. With the introduction of sodium tripolyphosphate (STP) as a dirt-sequestering agent in detergents, the amount of Phosphorus in the Great Lakes shot up. Algal growth increased dramatically each summer. In the autumn, the algae died and fell to the bottom of the lakes, where it decomposed, stripping oxygen from the water. The lack of oxygen resulted in the death of fish eggs and a declining fishery, among other things.

The solution was straightforward: we created regulations under the Canada Water Act to ban phosphorus-based detergents and required that sewage treatment plants install phosphorus removal systems.

Dealing with the other pollutants in domestic wastewaters is far more challenging—pharmaceuticals, toxic chemicals, plastic microbeads, antibiotics and hormone-mimicking chemicals are ever-changing and very complex to deal with. Public education about these hazards, along with a positive community spirit, may be the only way to prevent the release of such materials into domestic wastewaters until a technical fix can be found.

This is but the tip of the iceberg with respect to water quality problems. The visible problems are billions of tons of plastics and brightly coloured plumes of chemicals from untreated industrial waste stream that exist world-wide to this day. More insidious are the invisible contaminants that are increasing in our global aquatic environmental commons.

The hydrologic cycle has been the world's great water purification system for millennia. When the atmosphere is saturated with water vapour to the point where it can hold no more, it drops that water as rain, snow, or hail. That water is pretty clean. It may contain dissolved gases and a few particulates, but it is generally of a drinkable quality. Problems associated with acid rain caused by industrial emissions of sulfur dioxide dissolved in water vapour in the 20th century have largely been dealt with.

A great proportion of precipitation goes to nurture vegetation, the rest soaks into the ground. Water taken up by plants and animals is quickly released back to the atmosphere in pure form by transpiration (plants) or respiration (animals). Where there are pollutants on the ground, rainwater becomes contaminated. If those pollutants are natural, like plant debris and animal wastes, the water cleans itself as it percolates through the soil. Organic compounds break down with natural bacterial action and the products of decomposition are oxidized, leaving clean water to add to the underground aquifers and to feed streams and rivers.

As streams and rivers flow to the oceans, they pick up salts, solids, and various pollutants. Water vapour then evaporates from the surface of the oceans, leaving pollutants behind. This is how the hydrologic cycle returns relatively pure water to us daily.

Think of that box of water described a few pages earlier. Your portion of water, in a box 130 feet by 130 feet by 130 feet would not be enough to carry you through a lifetime of growing and preparing food, drinking, washing, fabricating goods, and so on. But the hydrologic cycle, if not abused, will return that box of water many times over in your lifetime, in purer form than you left it, provided it is not abused along the way.

My childhood farmer neighbour carefully balanced his dairy cattle feed and manure management to keep the environment clean and productive. Present day preoccupation with maximum profit now dominates, damaging the hydrologic cycle and removing large portions of water from human use for generations. It is considered more efficient to run "feedlot" operations to produce beef. Cattle are constrained in pens and fed feed which is trucked to the site. On the positive side, less land is required to fatten cattle on grains trucked to a feedlot operation than is the case for free-range cattle. On the negative side, the concentration of cattle waste in one place contaminates the air and water. Feedlot operations are situated in areas where there is sufficient water for the needs of thousands of cattle. Often this is above a suitable aquifer. Liquids from cattle manure are too concentrated for natural purification and they contain antibiotics and hormones that eventually percolate down to the aquifer, making it unsuitable for use. When the aquifer has been contaminated, the feedlot operation moves on.

This is a relatively minor example of interference in the hydrologic cycle by putting too much waste in one place. More serious are the problems that arise from the discharge of persistent toxic chemicals which cannot be neutralized by the normal hydrologic cycle. Such chemicals come from a wide variety of sources: chemical process facilities and other industrial plants; pesticide contaminated run-off from farms; drug and hormone contaminated effluent from sewage treatment plants; and run-off from urban and suburban roads, parking lots, and lawns. These wastes may be directly poisonous (lead, mercury, cobalt, and chromium), they may cause cancer or chromosome damage leading to death, or they may cause genetic changes that lead to birth defects. Persistent toxins can be organic, like PCBs, or inorganic like lead or mercury. My late wife, Anastasia Shkilnyk, wrote the remarkable book, *A Poison Stronger Than Love* (Yale University Press), clearly describing a mercury poisoning situation in Northern Ontario.

A couple of my earliest research projects examined how things like cadmium could interfere with metabolism, and how PCBs interfered with the reproductive cycle of gulls in the Great Lakes. The problem with these chemicals is that nature's hydrologic cycle cannot neutralize them, and they may contaminate water supplies forever. We need to do everything possible to prevent the introduction of such chemicals into the environment in the first place. Recalling the Niagara River anecdote with which I began this chapter, we must not allow commercial interests to poison our limited common water resource forever. Thinking of that 40-metre (130-foot) cube that represents your share of the world's fresh water, you can go on using it for a lifetime if you don't poison it.

Just because we *can* poison our own well, doesn't mean we *should*.

Land

"Buy land—they're not making it anymore."
(MARK TWAIN)

Well, Mark Twain got it partly right; there is only so much land to go around. Let's begin with a brief assessment of the global situation. There are about 150 million square kilometers of land on Earth. That's about 4.75 acres per person. (My apologies to purists for switching units of measurement here; most North Americans think about area in acres or square feet.) That is about one third larger than an American Major League baseball field. When you subtract the one-third of this land which is unsuitable for human use or habitation because it is desert, mountaintop, the Antarctic, and so on, you are left with an area about the size of that baseball field (roughly 400 feet by 400 feet) with about half of that area covered with forests, lakes, and shrubs.

Next, removing forested areas, shrubs, and lakes from your share of the world's useful land leaves you with less than half the area of a baseball field. That's approximately the area encompassed by the triangle from behind home plate to the left-hand corner of the field and across the back edge of the sandy bit behind second plate over to the right-hand corner, which is the shaded area in the diagram below.

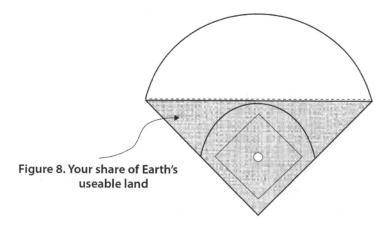

Figure 8. Your share of Earth's useable land

Imagine you are sitting high in the bleachers of a baseball stadium looking down on your land: it's not a lot of land when you consider that you need to put your home and garden there, leave enough space to produce all your food and still have room for your share of schools, hospitals, roads, parking lots, factories, offices, and shops. Producing your food takes up most of your patch of land; more than half of it, and the area available for agriculture, per person, is rapidly dwindling. In 1960, there were about three acres of farmland for each person on Earth; in another twenty years our exploding population will have reduced that to around 0.03 acres per person.

Most of the world's people live in urban areas (54% in 2014 and expected to grow to 68% by the middle of this century). Urban areas are estimated to total about 370 million acres. Growing food happens elsewhere. While that seems like a lot of land for cities and suburbs, it boils down to just 0.09 acres per urbanite. That's less than 4,000 square feet, and it will shrink to about 2,500 square feet per person by 2050.

So, let's return to that baseball field and try to visualize how much land is available for an individual's residence and that person's share of roads, parking lots, offices, factories, schools, hospitals, and so on. Imagine you are at home plate, and you walk twenty-seven paces towards first base. That's about two-thirds of the way there. Cross over to the pitcher's mound, across to the two-thirds mark on the third base line and back home. That's it. That's your share of land for all your urban needs. Less than the size of a large North American house.

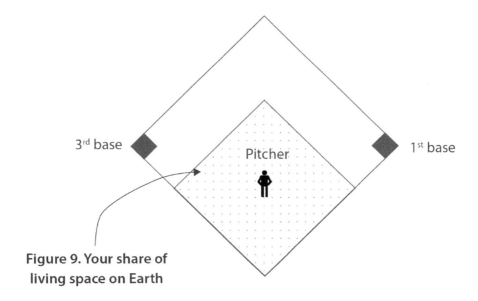

Figure 9. Your share of living space on Earth

No, that is not a misprint. If all land was distributed equally around the world, your *urban* share would be a square 63 feet by 63 feet (4,000 square feet, about 27 paces by 27). But land is not distributed equally around the world. North Americans in urban and suburban settings have a lot more than the world average. Southeast Asians have a lot less. The rest of the world is somewhere in between. One of the many reasons for increased migration is that people want to move from areas of severe overcrowding to places where there is more open space. One of the reasons for resistance to immigration is that people who now enjoy a bit of breathing space do not want to see that space diminished. Since we cannot create more land, these competing desires can only by accommodated by reducing or stopping the rate of growth of the population; ideally, holding population constant. Given today's rate of growth, at about 80,000 additional people each day, we have a lot of work to do to achieve equilibrium.

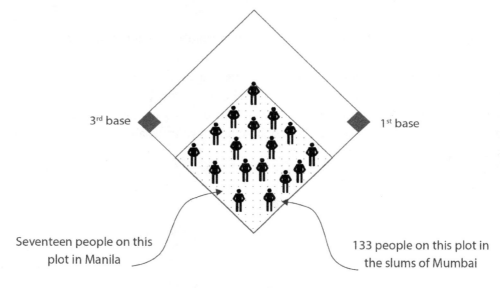

Figure 10.

Of course, people aren't spread evenly from city to city, or within cities. In Manila, there would be seventeen people, plus their share of roads, shops, offices, schools, and factories on that little plot of land. Their living space is just a fraction of that, so they must stack themselves on top of one another, in tenement buildings, to make room for the services needed. The population density of the slums of Mumbai would have ten people living on an area the size of the average living room in North America. Hardly a humane way to live.

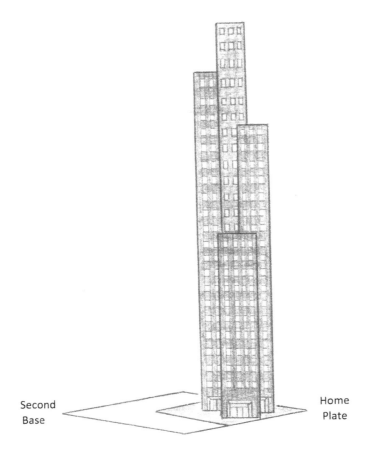

Figure 11. Stacked populations in cities

Would you really want to be one of thousands of people living on top of one another in a city like Manila?

We are not doing very well at holding the population constant, we are only just beginning to slow its growth rate. Considering this, we try to increase the amount of land available for human habitation and agriculture by means of deforestation. Deforestation of land for conversion to food production is not a new phenomenon. The Chinese began doing so centuries ago. Europeans followed suit, cutting down half their oak, beech, and birch trees, as did the Americans. Even the American Indians of the northeastern areas of what is now the United States burned swaths of land to create meadows, which were more useful for increasing the deer population, thus improving the quality of the food supply.

While those activities were generally not disruptive of the dynamic equilibrium that existed between humans and the environment in which they found themselves, the same cannot be said of the present deforestation of the Amazonian rainforest. Cutting trees and burning the debris adds directly to greenhouse gases, while rainforest destruction removes vast areas of forest that act as a natural CO_2 sink. At the current rate of deforestation in Brazil, the contribution to greenhouse gas buildup is more than the combined contribution of such gases from all the world's ships, cars, trains, and planes.

Because we have already converted most of the world's grasslands into feeding areas for livestock, we are now resorting to clearing more of the world's forested areas to provide more farmland for a growing population. It is hard to get accurate figures on the rate of deforestation, but a rate of about 15 million acres per year is most likely. To get a practical feel for that, let's go back to that baseball field. Imagine that every square foot of it is covered with a mighty forest. Now, close your eyes and count slowly to five. When you re-open your eyes, you see only a jumble of fallen logs, stumps, and brush. Every five seconds! Let's do that again—imagine the entire baseball field filled with a towering tropical forest teeming with life, butterflies and birds, mammals and insects, and so on. Close your eyes for five seconds and reopen them—to see a field of ash with a pall of acrid smoke hanging over it.

According to the United Nations Food and Agriculture Organization, the most significant driver of deforestation is the cutting down or burning of trees to make space for more agriculture. Deforestation for beef production causes more than twice the deforestation attributed to the production of lumber, soy and palm oil combined. High demand for meat products has led to the conversion of millions of acres of tropical forests into grazing lands.

While we are converting the world's forests into agricultural land, we are at the same time converting agricultural land into urban and industrial land. This is especially a problem in North America where, in the United States, for example, one and a quarter million acres of farmland are being removed for other uses each year. The United States has about three acres of food producing land per person now, with most of that used for grazing cattle. With population growth and the removal of land now used for food production, land available for food crops will drop to around three-quarters of an acre per person by the middle of this century.

Let's think a bit more about your personal 4,000 square foot plot of land. What are you going to do with it? Dump garbage on it? Build more roads and parking lots? Think carefully, because each individual plot is getting smaller day by day as the world's population grows.

Land is not distributed equally around the world. In many places, crowding has reached a level that is already inconceivable to most North Americans. New York City, with a population of about 18 million people living on about 8,700 square kilometers, ranks as only the 114th most densely populated urban area in the world, with about 5,000 square feet per person. As noted, most of that is taken up with roads, offices, shops, factories, schools, hospitals, and so on. Less than one-third of that area (about 1,600 square feet per person) is used for residences, taking account of both single-family residences and apartments. Mumbai, India, is about fifteen times more populous than New York City. In parts of Mumbai, an equal portion of land for each individual would measure less than ten paces by ten paces.

In many cities, the solution to population pressure on limited land is to build upwards. Thus, the space per person on a given patch of land increases ten-fold in a ten-story apartment. Clearly, such a strategy has limits—we won't be building thousand-story apartments any time soon. In any event, that would not be a sensible way to live.

So, to summarise, in the time it has taken to read this section about land, your living space has diminished significantly and sixty baseball fields worth of forest have been destroyed.

Why does this matter?

This matters because, since the beginning of life on this planet, there has been an equilibrium between plant life and animal life that has been maintained to the benefit of both. Sunlight bathes plants with solar energy. In the leafy parts of plants, tiny plant factories, chloroplasts, use this solar energy, along with water and CO_2, to make carbohydrates and oxygen. In animal cells there are mitochondria, tiny animal factories that use those carbohydrates and oxygen to make energy as they release water and CO_2; this is the energy that keeps animals warm and allows them to move about. A perfect natural equilibrium: plants use water, CO_2 and energy from the sun to make carbohydrates and oxygen, animals use those carbohydrates and oxygen to produce energy (which allows them to do things) and water.

For millennia, these systems have been in equilibrium. In our time, the Anthropocene, as some call it, we are burning carbon at a frantic rate, carbon that took millions of years to store, and increasing the levels of CO_2 in the air, while at the same time removing forests at a frantic pace, thus removing nature's way of using up CO_2 to produce oxygen. It is called the Anthropocene because it is the period during which human activity is shaping the environment. (Anthropocene means "the new, *human*, era")

As we destroy forests, especially the lush, diverse tropical forests, we diminish the Earth's ability to soak up carbon emissions and produce oxygen. In 2018, a typical year over the last decade, another 12 million hectares of forests were lost. Almost 3 million Hectares of those losses resulted from wildfires around the world and approximately 3.6 million hectares of those losses were primary rainforests. That's about 9 million acres of primary rainforest destroyed; an area larger than Belgium.

We are in serious trouble in this, the third component of the Global Environmental Commons—I quite a tragedy.

Just because we *can* lay waste to the land doesn't mean we *should*.

Species

EX HAC TELLURE MERITA NOSTRA CURAE ET IN OMNE TEMPUS,
IN COMMODIS OMNES SPECIES.

2,050 years ago, Cicero wrote these words, which mean: *This planet merits our care for all time and for all species.*

Over the last half a billion years there have been five major extinction events.

The first major extinction, the Ordovician-Silurian extinction, occurred about 440 million years ago, and saw the loss of about 85% of life on Earth. The Earth's atmosphere cooled, perhaps because of increased volcanic activity, growing glaciers, and falling sea levels. Dropping sea levels may also have been caused by the growth of new landforms like the Appalachian mountain range. This extinction event was spread over about seven million years.

In the longest but least severe of the five major extinction events, the Devonian extinction saw the loss of about 75% of animal species over a span of about 20 million years (380 to 360 million years ago). The cause of this mass extinction event is unclear.

The worst mass extinction, lasting about 60,000 years, was the Permian-Triassic extinction, which saw the loss of about 96% of all species about 250 million years ago. Its causes are unclear; increased volcanic activity and/or a meteor impact may have been the cause. Humans and all other life forms we see around us evolved from the 4% of species that survived the Permian-Triassic extinction.

Around 200 million years ago, the Triassic-Jurassic extinction extended over about 10,000 years and resulted in the loss of about three-quarters of the species then alive. The most accepted cause of this extinction is extensive volcanic activity that released large quantities of carbon dioxide, methane, and sulphur.

The last of the five great extinctions was the Cretaceous–Paleogene extinction, about 65 million years ago. It is thought to have been caused by an immense meteor impact that ejected such large clouds of dust into the atmosphere that sunlight could not penetrate to the surface of the Earth in the normal way, resulting in the death of about 75% of the species alive at the time.

These five major extinctions have a couple of things in common. One is that they occurred over long time periods of thousands to millions of years. The other is that they were caused by events not under the control of the life forms that lived on the planet at the time of the extinction event. The sixth major extinction event is happening now.

This time is different.

At the present pace, this extinction event will be more rapid than previous events and will be caused by a single species living on Earth now—Human Beings.

In a very much simplified graphic form, the change in the number of species on the planet since life began looks something like this:

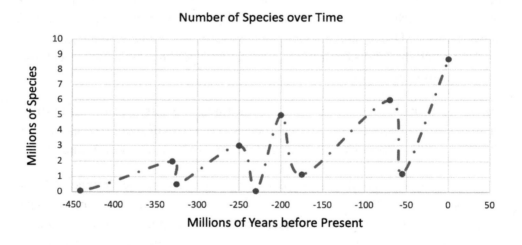

Figure 12. Number of species over time

While it is reasonable to expect a higher extinction rate over the last hundred years because of human activity, it is impossible to accurately determine the rate of extinction. This is because we do not know how many species of plants and animals there are on Earth, and we do not have the ability to accurately measure the number lost each year. Scientists estimate that there are around 8.7 million species, but they have identified and catalogued only a small fraction of these. It is equally difficult to assess the rate of extinction of species. It is reasonable to conclude that many are lost because of deforestation, pesticide use, hunting activity, and climate change, but for every disappearance, we notice there may be five or ten species we didn't even know existed that have disappeared along with them.

Taking the average of estimates from studies done by different groups of scientists using data collected from around the world, both on land and in the water, it is not unreasonable to assume that we are losing about twenty species every day. At this rate about 8% of the world's species will become extinct over the next one hundred years.

Looking at the number of animals that have disappeared, as opposed to the number of species, it appears that 58% of higher life forms (birds, fish, reptiles, amphibians, and mammals) were eliminated in the forty-two years between 1970 and 2012. Populations of animals living on land appear to have declined by 35%, with ocean life declining by 36% and freshwater animals by 81% over the same period. This matters because we depend on many of the life forms that are now disappearing. Unfortunately, given the remarkable inter-connectedness of life forms, we may be far more dependent on other disappearing animals and plants, whose loss we have yet to tally, than we realize.

One obvious dependency is our relationship with pollinating insects. Bees pollinate about seventy crop species that feed 90% of the human population. Other animals also depend on plants that bees and other insects pollinate, and we may depend on those animals for our own nourishment, so there is a compounding effect here. There has been a serious decline in the numbers of pollinating insects, especially bees, over the last few decades.

Another obvious dependency is our long-standing reliance on seafood for sustenance in many parts of the world. Properly managed, wild populations of fish and other marine animals could provide vast quantities of food for humans forever. In the not too distant past, humans lacked the ability to fully exploit the marine resources available to then, let alone over-exploit those resources. Now, only about 10% of marine stocks are less than fully exploited. We now fully exploit 60% of fishery stocks, and the proportion of total fish stocks that are over-fished grew from 10% to 30% in the thirty-five years between 1974 and 2009. Clearly, the situation with respect to over-fishing is deteriorating. As will made clear in the Food chapter in Part II of this book, this is not an accident.

As fish stocks that used to be readily available to coastal fishing nations are depleted, fishermen travel beyond the great fisheries of the North Atlantic and the North Pacific to more southerly oceans, and they journey much further offshore to the deeper waters of the ocean. A few decades ago, fishermen struggled to catch fish down to 500 meters below the surface; now they are scooping up species at depths of 2,000 meters. Here again, as we deplete species most valuable and sustainably useful to us, we move on to more remote, less valuable, species and over-exploit them as well.

The same sort of thing is happening to freshwater fisheries. The Yangzi River in China offers a case in point. Fishermen have been harvesting fish there for many centuries. To feed a burgeoning population, the catch was increased to the point of being unsustainable. At the same time, the need for energy grew dramatically, with the result that the river was dammed, further damaging fishery resources there. The catastrophic declines in fish population made it necessary for the authorities to close fisheries, putting 280,000 registered fishermen out of business. To make sure they stay out of business, their boats have been smashed and hauled away. So, the present generation no longer has a supply of local fish to provide their protein needs, at least 280,000 families have lost their major source of food and income, and some species of Yangzi fish have been driven to the point of extinction. Future families may be forgiven should they curse their ancestors for their lack of foresight.

The most offensive destruction of animals is that which occurs because some humans like to exhibit them as trophies or use animal parts as trinkets or for mythical health benefits. Somehow, some men still believe that roaring around the jungle in heavy all-terrain vehicles and shooting elephants with large calibre "Elephant Guns" is a demonstration of their manhood. Others believe that the display of ivory trinkets demonstrates how wealthy they are. Still others believe that by eating the equivalent of hair or fingernails (the kind of material of which Rhinoceros horn is made) they will be made more virile. If only they could conceive of a topsy-turvy world in which the elephant or the rhinoceros was king, and humans were killed so that trinkets could be made from human teeth or potions from human fingernails: their attitudes might quickly change.

Here are some of the "Endangered" and "Critically Endangered" species of large animals:

ENDANGERED SPECIES

African Wild Dog	Humphead Wrasse
Amur Tiger	Indian Elephant
Asian Elephant	Indochinese Tiger
Bengal Tiger	Indus River Dolphin

Black-footed Ferret	Irrawaddy Dolphin
Blue Whale	Mountain Gorilla
Bluefin Tuna	North Atlantic Right Whale
Bonobo	Red Panda
Borneo Pygmy Elephant	Sea Lions
Chimpanzee	Sei Whale
Fin Whale	Sri Lankan Elephant
Galápagos Penguin	Tiger
Ganges River Dolphin	Whales
Green Turtle	Whale Shark
Hector's Dolphin	

CRITICALLY ENDANGERED SPECIES

Amur Leopard	Saola
Black Rhino	Sumatran Tiger
Bornean Orangutan	South China Tiger
Cross River Gorilla	Sumatran Elephant
Eastern Lowland Gorilla	Sumatran Orangutan
Hawksbill Turtle	Sumatran Rhino
Javan Rhino	Vaquita
Malayan Tiger	Western Lowland Gorilla
Orangutan	Yangtze Finless Porpoise

So, there you have it: the "Tragedy of the *Global* Commons."

Over the last five thousand years, and the last two hundred years in particular, we have managed to pollute the air we breathe so badly that each year about eight million people die prematurely because of air pollution. We have discharged so much warming waste into the air that the entire planet is heating up.

Your personal lifetime box of water, roughly the area of a baseball field covered to a depth of twelve feet, is getting smaller every day and is becoming permanently contaminated with persistent toxic chemicals. That's your share of useable water for life, for home and garden needs, for laundry, cooking, swimming, health care, car washing, lawn watering—everything.

Your personal piece of urban land, that square about 27 paces by 27 paces on which sits your living space and your share of roads, hospitals, factories, shops, and so on, is shrinking day by day. In the time it takes to read this small paragraph, another two or three baseball-field-sized chunks of forest have disappeared.

While you know that species and the number of plants and animals is in precipitous decline, you cannot know the ultimate effect of the loss of one or more of those species on your own survival.

How did we get into this mess?

Part II
WASTED BY DESIGN

The first part of this book examined the state of *The Global Environmental Commons*, the air, water, land, and species on which we depend for survival. The wounds mankind has inflicted upon the environment run deep and are found everywhere. To better understand the way to heal these wounds it will be helpful to understand how we created them in the first place; that is the purpose of this part of the book.

Part II, *Wasted by Design*, looks at some of the many ways we have deliberately and knowingly injured our environment. It begins with the chapters "Everywhere you Look" and "What You Don't See," which describe a few of the millions of things we encounter each day that are harmful to the environment. Then several areas of human activity are examined to see how conscious and sub-conscious actions have been harmful. These areas are addressed in order of their importance to our daily survival: Food, Shelter, Security, Health, Energy, Transportation, Durables, and Recreation.

Everywhere You Look

Imagine reading the thoughts of a typical North American male as he goes through his working day. You might read thoughts like these.

00h00-24h00: From the first second of every day to the last, there are dozens of little LED lights showing that various devices are connected to the electrical grid. There are digital clocks telling the time, whether anyone is there to look at them or not. There are mini-incandescent lights on power bar switches. Because the power consumption for LEDs seems quite low, we tend to dismiss them as a trivial drain on the environment. But a typical microwave uses more electricity to power its digital clock each year than it uses to heat food. That's because most microwave ovens stand idle for more than 99% of the time—but their clocks are still running and consuming power. The same thing happens in modern ovens, clock radios, television receiver boxes, night light switches, and so on. Walking through the house in the middle of the night with the lights off reveals little lights glowing in every room. Each one represents a deliberate waste of energy.

06h30: Time to shower. A daily routine. Think about that—our parents didn't shower every day. Our grandparents probably grew up with one bath each week when they were young. They seemed to survive quite well, and I don't recall that they had unpleasant odours. Now, it seems, everybody has a daily hot shower, using plenty of soap and shampoo, followed by a generous application of synthetic deodorants and anti-perspirants.

06h45: Now to shave. There was a time when men used straight razors—neither safe nor comfortable. Then there was the safety razor with a reusable metal handle that lasted for years—only the lightweight double-sided blade was thrown out. They didn't recycle things in those days. Then came the cartridge blade with its combination of plastic and steel—*that* wouldn't be recyclable anyway. Now we have the disposable razor; handle and head with twin (or triple, or quadruple) blades. One just throws the whole thing away after a few days; it can't be recycled. Soap pumps. Too bad refills are unavailable. Oh well, the ten packs of pumps are cheap, and one can just throw away the empty glass bottle and its pump assembly with its metal spring and non-recyclable plastic parts when empty. That will keep the manufacturers humming.

07h00: What to wear? So many choices. People don't want to be seen wearing the same outfit to the office day after day. Better to have a different outfit for each day of the month. How many acres of land went into the growing of the cotton for these outfits? How many barrels of oil went into the production of the synthetics? Who cares?

07h15: Put the kettle on for coffee. Which coffee pod do I feel like today? The Columbian, the dark roast, the breakfast blend, perhaps? There must be twenty types to choose from. Where do all those pods end up? About 170 million are thrown away each day. The kettle, filled with cold water, is taking a very long time to boil. Most of the contents won't make it to the coffee cup—the extra will be poured down the sink.

07h30: Another traffic jam. It seems this happens almost every day. Why are there so many people on the roads these days? Why don't they take public transit? Why doesn't everybody take public transit?

08h00: Pick up a ready-made breakfast from a drive-through establishment. Notice how much wrapping is used and then dumped in the nearest trash can. I wonder how many disposable cups they go through each day. Very convenient though, one can eat and drive at the same time.

08h45: Straight into the office paid-parking lot. A few acres of wasted land in the most valuable real estate of the city.

09h30: Early coffee today in advance of a Coast-to-Coast Directors meeting, there are dozens of different flavours of coffee pods to choose from. Even more waste than at home.

10h00: Everybody is here, from across the country. Good thing there are plenty of flights to choose from, from the east coast and the west, from north and south. Good to see the guys in person. It may be expensive, but who cares, at least there is no carbon tax on air flights. Joe from the east coast has this great idea on how we can increase profits by 2%, just by putting 2% less product in our boxes. The customer will never notice. It means an additional 2% unnecessary packaging waste, but that's not *our* problem.

12h30: Lunch with our colleagues at a favourite restaurant. They sure love their steak, especially when someone else is paying. Altogether, they must have ordered half a dozen sixteen-ounce steaks and this place sure knows how to heap the plates with the side orders.

14h00: I have to print off another twenty copies of a colour report for this afternoon's meeting. The printer says, "Cartridge low on ink." Where are those cartridges? Can't use the ones from the old printer. Same model, but the new model has a slightly different cartridge shape and the old ones don't fit. Staff bought a case load of each colour just before the old machine packed it in. They don't make the old model anymore, so all those cartridges will just be thrown out. Why don't they sell ink by the bottle so one can just refill the cartridges up when they get low? Because they have to increase profits, just like we do.

18h30: It's been another long day, and here I am stuck in another traffic jam on a sweltering afternoon. Why don't they build more freeways in this city? Look at all the time wasted with everyone stuck in traffic.

19h30: Finally, a lovely supper. My wife found a great deal on New Zealand lamb, and she put some Kiwi fruit along with some wedges of orange for a tropical fruit salad dessert. When I was a kid, we only had apples, pears, plums, cherries, grapes, and raspberries for desserts. Now we can have anything we want from anywhere in the world. We even have bananas from the tropics all year round. Ever wonder how many bananas are shipped each year? (Answer—more than 100 billion, about twenty million tonnes.)

20h00: Great discussion over dessert about this year's winter holiday. We had such a great time looking at the pyramids in Egypt last year, we think we might do two trips this year. A combination trip to Laos and Cambodia seems like a good option. Watching the Buddhist monks parading down the main street of a quiet Laotian village; seeing the great temples of Angkor Wat. We'd have some great stories to impress our friends with when we get back home.

20h30: Time for a little TV. Some action-packed thrillers, I guess. That's about all they show these days. Car crashes, buildings exploding, bodies spurting blood as the bullets zip through. I wonder what effect these shows have on the kids these days. Oh well, what can you do about it? That's what sells the ads, and the ads keep the economy humming, and that's all that matters. Look at those ads—how they glorify waste! Whether it is driving a car to the top of mountains, over desert sand dunes, or through forest streams, you've just got to have one of those muscle-crossover vehicles to get to the office or the corner store. There are ads encouraging kids to waste as much toilet paper as they can—there is always plenty available. And on, and on, and on…

23h00: Time to turn in. Good thing we have air conditioning. With all the night-time noise, I don't think I could get to sleep with the windows open.

24h00: Yep—all the digital clocks and LED power lights are still burning.

These are some of the fictional daily thoughts of a typical office worker. Yours will be different. There are probably more like a million different things we see each day that illustrate how the planet's environment is wasted by design. Every person's observations will be different and will change somewhat from day to day.

Everywhere you look there are examples of how everyday human actions deliberately abuse the world's environment and its limited resources. These examples surround us, but we often fail to see them for what they are. Sometimes this is because the unintended environmental consequences of following the directions of business leaders, political leaders, financial gurus, religious leaders, and even our neighbours, are deliberately obscured from our view. Too often it is because we accept what is offered to us in an uncritical, unquestioning fashion. We are preoccupied with the present: the present business cycle, looking only three months ahead; the present political cycle, limited to about four years; the present human reproductive cycle of about twenty to thirty years; the present life cycle of perhaps seventy-five years. We are reluctant to have our gratification within these cycles jeopardized by altruistic considerations concerning the quality of the environment we leave for those who may follow, five or more generations down the road.

The environmental damage that results may take hundreds of years to be fully apparent, and it may last for hundreds, if not thousands, of years. These are time frames we are reluctant to embrace. The trouble is, most of us don't even think about the environment when we see these things. That is the first thing that must change if we are to find solutions for this wounded planet.

Things That Are Hidden From View

So often ways that the environment is deliberately wasted are hard to recognize or deliberately hidden so that they can't be seen. Trivial things like the rubber washers in your garden hose that are made in colours close to the colour of grass or earth so that if one falls out while changing fixtures, you'll probably never find it. Complex things like a small defect in one part of an appliance that requires the replacement of an entire machine assembly because the individual defective part cannot be purchased.

To illustrate, here are a few of the often-hidden things that waste the environment by design:

1. Refrigerator-freezers that automatically defrost twice a day whether there is an ice build-up or not. A great deal of energy is wasted every time this happens, first to heat the ice compartment then to re-cool and refreeze the system. At the very least, there should be an option for the user to adjust the cycle to meet the need. A better option would be to require that all such appliances be equipped with a frost sensor to determine when and how much defrost is needed.

2. Appliances like stoves, ovens, dishwashers, and clothes washers equipped with complicated interdependent electronic circuits. When something goes wrong it is often more costly to repair than replace. Sometimes replacement parts are just not available after a few years. The manufacturers would prefer that the consumer purchase a new unit rather than repair the old. Is it so demanding for a cook to turn one knob to set the temperature and another to set the length of time to cook? In my experience, it takes longer to set all the digital computerised devices than was the case for the older models.

3. Food is now almost always packaged in plastic. Plastic clamshell containers for berries and small tomatoes, Styrofoam wrapped with cellophane for meat and many vegetables. Lettuce wrapped in cellophane. At the checkout, all is put in plastic bags to take home. In some places every apple, every pear, every banana or avocado carries its own label. This has become so universal that we

have come to believe that there is no other way. Just a few decades ago there was no such overwrapping. Every day billions of plastic containers are made, sold, used once and thrown away. Despite good intentions, most wind up in the environment, often in the ocean. It may make things more profitable for industry but, ultimately it costs the consumer—and the environment. There are alternatives—consider string bags that can carry many loads of groceries for many years.

4. Every year we celebrate holidays like Easter, Mother's Day, Father's Day, Halloween, Thanksgiving, and Christmas. There is nothing wrong with celebrating special events. But when the celebrations become a contest to see who can assemble the most shocking Halloween display or the most glorious, energy-consuming Christmas display, we have moved into the realm of excessive environmental waste. The TV channels glorify those who have made the most ostentatious display, whatever the holiday. So, your neighbour hopes that, this year, he might be recognized for the most outrageous waste of all. Think of all the energy and resources that went into making the display, only to be thrown away afterwards. Often such events are also associated with gross overeating, with its associated bad health effects and an environmental hangover of food scraps thrown into a garbage can.

5. Thinking of food and its environmental impact, food production has been globalized, and agribusiness has become profitable as it provides goods from around the world. Part of the profitability derives from the absence of accounting for the sum of all the environmental costs associated with the remote production and international trafficking of food products. Although local producers provide varied and bountiful food, shoppers still seek exotic goods from foreign lands. Environmental damage associated with our hunger for foreign foods ranges from excessive clearing of forested lands and loss of local biodiversity on those lands, to the consumption of non-renewable fuels and the emissions associated with the combustion of those fuels. Such environmental costs are not figured into the cost of food in the developed importing countries. Even simple goods like pickles are exported from South-east Asia to North America. These are heavy and of little intrinsic value, but have a high environmental cost associated with shipping them. There is an added social/environmental cost associated with the underpayment of labour in the producing country for the food provided and the diversion of local labour from less environmentally-damaging food production practices. If all these costs were properly accounted for and included in the price of imported food, the consumer might choose to purchase locally grown food, especially at harvest time.

6. Consumer goods are often sold in multi-option packages even though only one option will ultimately be used. Think of a box of solar-powered LED lights for outdoor use. And they come with mounting hardware for all kinds of applications. One can put them on a ledge, mount them on top of a stake, hang them on wall hangers, hang them on hooks that attach to posts and stakes... Once installed in the purchaser's preferred option, the rest of the mounting paraphernalia is thrown away. The deliberate waste is seldom, if ever, considered.

7. Hotels will often offer, or insist upon providing, bottled water in their restaurants even though their promotional literature makes it quite clear that the hotel's tap water is safe to drink. It may well be that the tap water is healthier than the bottled water, yet customers seldom ask for the healthier and environmentally less damaging option.

8. Many consumer goods, from sugar, to fertilizer, to nuts and bolts are sold in prepackaged quantities—quantities that may exceed the customer's needs, or be insufficient, thus requiring the purchase of several units of the prepackaged items. Pre-packaged food that exceeds the consumer's normal use pattern is wasted. Packaging is sometimes multi-layered, with a plastic bubble glued to a cardboard backing and sometimes encased in another layer of plastic. This makes reuse or recycling very difficult, if not impossible, and often goes unnoticed by the casual purchaser.

9. Public transit systems are often poorly designed, with more attention devoted to up-front capital costs than long-term operating costs and the pollution associated with inefficient operation. Often, all buses are the same size on all runs at all times even though it is known in advance that they will often run at less than full capacity. Too often they operate at less than 10% capacity. In many areas, schedules are very light. People decide to take the alternative, personal car transport, and the schedules get lighter, costs increase, fares increase, and people use the service less.

10. Many devices, especially technological ones, are made to fail with a flaw in an almost insignificant and cheap part. That flaw will result in rendering the device useless. Here are just three of the countess examples I have seen recently:

    i. A computer mouse stops responding to the most used, left-side clicker. The local tech shop advises that the mouse is now worn out and needs to be replaced—it apparently can't be fixed. Taking apart the mouse and examining the little bumps on the underside of the clicker, reveals that the bump meant to push the switch inside has worn down.

    ii. The microwave dial that sets the time to cook spins uselessly. Getting inside the device is made very difficult so the owner can't get in to tighten the nut that has come loose inside. The local appliance store advises it will cost more to fix than replace.

iii. The neighbour's leaf blower suddenly starts to make an exceptionally loud noise. Probably something stuck in the fan blades. But the blower is assembled with twelve screws, two of which have unique heads, making it impossible for the handyman to disassemble. The garden supply store advises that it is time to buy a new blower.

In each of these cases, there was a possible fast and inexpensive fix; it's just that the manufacturers don't want you to fix, they want you to replace.

11. You can buy floor cleaners and toilet bowl cleaners that will apparently kill 99.9% of bacteria on your floors and in your toilets. Apart from the obvious question of why you would want to do that when the next time you walk across the floor or use the toilet it will again be back to normal, one should think of the bacteria that are not killed. They could survive to start a new strain of cleaner-resistant bacteria that would be even more difficult to remove when it is really necessary to do so—on eating utensils or surgical instruments, for example.

12. Often the cost of replacement batteries for old cordless power tools cost more than a new tool that serves the same purpose. Consider the cordless drill. When the rechargeable batteries on the old drill can no longer hold a charge, the manufacturer will charge more for a set of replacement batteries than the cost of a newer model tool with its charger and carrying case. The old model, still in good order except for the batteries, is to be thrown away in its entirety.

13. All too often minor parts in a more complex assembly will not be made available by the manufacturer. A critical part of a car door handle, perhaps worth one dollar, cannot be bought. The entire assembly, costing hundreds of dollars, is the only option. The remainder of the old handle assembly is to be thrown away.

14. Millions of people in overpopulated parts of the world depend on charitable donations from abroad for survival. With four, six, or eight children, they just keep on having babies, expecting that their God or foreign donors will provide. Neither their God, nor the foreign donors help them towards a path of equilibrium dependent on a stable population. To the contrary, they are encouraged to have more babies in order to be good adherents to their religious group. Foreign aid agencies are discouraged from providing any advice on family planning or birth control.

15. What passes for "entertainment" is blatant environmental waste. Apparently, people like to watch this deliberate waste, so much so that billions of dollars

will be spent to glorify it. Sometimes the waste is obvious. Pie-throwing contests are one example. The "great" Tomatina tomato throwing festival in Valencia, Spain is another. In that spectacle, 150,000 tomatoes (40 tons) are thrown at people, to the delight of television viewers around the world. Then there are food eating contests where the challenge is to see who can stuff the most hamburgers down his throat in a short time.

16. There has been a proliferation of unnecessary technology everywhere, from telephones to radios, television, and cars, among other things. The telephone used to be quite reliable. But then it had to be made wireless, mobile, digitized, video-compatible, text-transmitting, and so on. With each new generation came a host of new problems, from service providers to format changes. Many feel they must change devices just to be in with the crowd. Same thing with television sets. It seems that they need to be changed every few years. Nothing is made to last anymore.

17. Automobiles are marketed for their ability to climb mountains, ford streams, and plow across sand dunes. The implication is that a personal mode of transport should be chosen for its power and accessories rather than its efficiency in moving people about with the minimum environmental impact.

18. Personal recreation activities have become power hungry and noisy. Dirt bikes, all-terrain vehicles, snowmobiles, and high-speed powerboats are examples of people "getting out in the environment" in the most environmentally damaging ways. The noise of a "cigarette boat" (a subset of speedboats), for example, precedes it by a few kilometers, to the dismay of everybody and everything within the sound envelope. Even underwater species suffer. All these devices destroy the quiet enjoyment of the environment by others. Their consumption of non-renewable resources and emissions of global-warming gases add insult to injury.

19. Advertisers in all media play up the joy of doing expensive things that destroy the environment with no social conscience about the damage done. This creates in the mind of the consumer the idea that doing such things is good, something to be desired and purchased. Consider an advertisement for the purchase of a lottery ticket for example. Inevitably, the winner spends his windfall in the most environmentally destructive way. The subliminal message is that we should all strive to have more money to do such things. There is no counter message about the damage done.

20. Governments continue to try to outdo one another in various space races. First to put a satellite in orbit. First to put a man in space. First to put a man on the moon, a man on Mars, a satellite on an asteroid, and so on. It is seldom,

if ever, explained that every taxpayer is paying a significant amount of money for each these ventures. Science commentators speak enthusiastically about how putting a satellite on an asteroid might help us understand the origins of our solar system. Knowing this, we will…? I suspect that the average taxpayer might oppose such glory races if he was told that for just a few billion dollars of his and other taxpayer's money he will be able to say with some certainty that the moon is not composed of asteroids, for example. In purchases we make individually, we can at least calculate the cost/benefit ratio. Not so in the space race.

This is not to suggest that all space endeavours are unhelpful. Communications satellites, weather satellites, and other earth monitoring satellites can demonstrate a reasonable cost/benefit ratio.

Now we turn our attention to how actions such as those described in this chapter play out in the various aspects of human activity and the portions of the environment most affected by these activities.

Food

HUNGER PANGS ARE MALADAPTIVE

For half the people of the world, hunger pangs lead them to seek instant satisfaction and gratification, which they readily find in foods full of sugar, salt, and fat; foods that make them fat and sick, and likely shorten their lives. For the other half, people respond to hunger pangs by seeking nourishment that is not there; they often go to bed hungry, and sometimes starve to death.

Hunger pangs evolved to let our ancestors know it was time hunt for a few more grubs and gather some nuts to eat. They ignored those hunger pangs at their peril. If they waited too long to begin their hunter-gatherer work, they could become too weak to go on. In some cases, even if they responded, their hunt would be unsuccessful, and they would starve to death.

In the modern era, hunger pangs are maladaptive because, in most of the world, junk food is readily available in great supply, 24 hours a day, every day of the year. Modern homes are sometimes stuffed with enough calories to enable the occupants to survive for months. Much of that food is simply unhealthy, packed with "empty" calories, fat, and salt.

Unfortunately, in today's world very few people give any thought to having a vegetable garden, let alone tending it organically. Instead of gardens, there are broad expanses of pesticide-laden, fertilizer-gobbling lawns. Many families reject available natural foods in favour of commercial products.

Producing food now takes up about 40% of the world's total land area, with roughly two-thirds of that for grazing livestock or providing fodder for livestock, and the remainder for vegetable crops for humans. Agricultural lands are under constant threat from urban, industrial, commercial, and other encroachment. With the human population near eight billion and growing, the demand for more food increases while the area of land available to produce it diminishes. One of the ways to fill the gap is through deforestation, which is occurring around the world in order to increase the amount of land available for food production. Such deforestation carries with it serious negative effects for the environment. The destruction of tropical rainforests in the Amazon, the Congo basin and elsewhere to make way for cattle farms and palm oil or sugar cane plantations, is particularly offensive.

The human body needs a variety of essential food inputs: *calories* for energy; several essential *amino acids* for protein production; a few *fatty acids* that the body needs but cannot produce itself; *minerals* for building bones and so on; and *vitamins* to help keep our cells functioning properly. How much of these nutrients we need depends on where we are in life's journey. The young growing body needs more amino acids as it builds more body tissue. The teenaged body needs a lot of energy if it is reasonably active (and a lot less if it is lazy). Adults in their prime may need more calories if they work in physically demanding jobs. Elderly people need fewer calories, but must watch their mineral intake to ensure they don't suffer bone loss.

For males, the need for calories peaks at around 2,800 calories per day for an active twenty-year-old and then declines over the decades to 2,200 calories per day when he reaches 70. The corresponding numbers for a male who is relatively *in*active are 2,600 calories per day at 20, declining to 2,000 per day at 70. For females the need for calories peaks at around 2,200 calories per day for an active 20-year-old and then declines over the decades to 1,800 calories per day when she reaches 70. The corresponding numbers for a female who is relatively inactive are 2,000 calories per day at 20, declining to 1,600 per day at 70.

Taking a rough average for both sexes, active and inactive, in mid-life suggests that 2,200 calories per day would be a reasonable energy intake. As it is, the average energy intake for American adults is more than 3,600 calories per day; 64% more than required. No wonder, then, that North Americans are generally overweight or obese. That 64% of deliberate overeating is simply waste. The food industry stimulates such waste by producing inexpensive, readily available prepared foods that are high in salt, sugar, and fat to entice the buying population. Europeans suffer from the same sort of overeating and food waste, being just moderately below the American average. If everyone ate no more than 2,000 calories per day and the food thus available were properly distributed, there would be no undernourished people (versus about 350 million undernourished now) and there would be a surplus.

At the opposite end of the scale, countries like Haiti, Ethiopia, and Eritrea try to survive on an average daily intake well below 2,000 calories per day; not enough to maintain a healthy body weight.

Because essential amino acids are common in animal-based foods, we need to assess just how much meat and dairy products are necessary to provide essential amino acids in a typical western daily diet. The data suggest that two ounces of pure protein per day is adequate for a healthy adult body. If we got all that protein in meat alone, that would mean little more than six ounces of meat a day. But a normal diet includes a variety of other foods that also contain protein. Dairy products, nuts, and beans, for example. We typically get about 40% of our protein in meat. That means we need eat only *about 2½ to 3 ounces of meat a day* in our breakfast, lunch, and dinner. Next time see someone wolfing down a sixteen-ounce steak, think about it as eating almost one week's worth of a healthy meat protein in a single meal.

The consumption of meat in the United States, averaged over the entire population, is about ten ounces a day. Given the number of children under five years old in the US, one can estimate the average daily meat intake of those over five at almost eleven ounces a day; that is about three to four times the amount needed. Overeating leads to health problems; it is also very damaging to the environment in terms of land use, atmospheric emissions, pesticides, and fertilizer waste. If North Americans are concerned about the cost of dealing with greenhouse gases through things such as a CO_2 disposal fee, they can more than offset that cost just by eating more sensibly. Even poorer countries like China are seeing a spike in meat consumption. The average person in China eats about half as much meat as the average American, but Chinese beef imports in 2018 were forty times what they were in 2010; a forty-fold increase in just eight years. Growth in beef production causes increased forest clearing around the world, and cattle rearing greatly increases emissions of greenhouse gases.

To add insult to injury, many people seem to revel in the sight of food being wasted for entertainment. I'm thinking here of events such the one in which an American, Joey Chestnut, eats 69 hot dogs and buns in ten minutes. Another example is Molly Schuyler, eating 501 chicken wings in half an hour at a "Wing Bowl" competition, or her feast of 13½ pounds of prime rib in a red meat-eating competition—with crowds cheering her on. That's the amount of meat that would sustain her in a healthy condition for more than eighty-four days, and she ate it in half an hour. What message does this send? This deliberate waste of food, and the environmental damage associated with its production, is rightly disgusting to the starving masses around the world.

The waste associated with food-destruction fun festivals is even more obscene. Tomatina festivals are now held in many wealthy countries. The original, and most famous in Brunol, Spain, prides itself on being able to destroy 150,000 tomatoes in a single day. Think of the time, land, fertilizer, and pesticides used and then wasted to create a huge mess, "just for fun." The residents of Ibi, Spain prefer to make a monumental mess while wasting tonnes of flour and thousands of eggs each year. Custard pie fights (250 people throwing 650 pies at each other in just a few minutes in Colchester, England, for example), and meringue throwing fights… The list goes on and on, and the media celebrate such events instead of drawing attention to how these events represent such a terrible waste of the environment.

Producing red meat as a protein source for humans takes a terrible toll on the environment. According to the United Nations Food and Agricultural Organization (the FAO), the livestock sector of food production is responsible for putting more than 7 gigatons of CO_2 equivalent into the atmosphere each year. (That would be more than 15,400,000,000,000 pounds of carbon dioxide, in case you were wondering. Sometimes our use of terms like kiloton, megaton, and gigaton obscure the size of the problem) This livestock sector of human activity, taking account of everything involved from converting natural habitats into pasture, through to processing and transport of livestock products, contributes almost 15% of human-activity greenhouse gases.

The environmental cost for the "transport of livestock products" mentioned in the previous sentence has a counterpart on the vegetable side of our diet, too. Without thinking about, it we expect to eat exotic fruits and vegetables from around the world any time we want. Bananas from the tropics are cheap; avocados appear magically on our grocery shelves. We buy these without ever thinking about how much energy was used to transport them to where we are. We ship our locally grown apples to faraway places while buying oranges abroad. It was not always so: the ancients ate only what could be found within a small radius of where they lived. This is not a plea to return to those dull old days, but rather a call to be aware that there is an environmental cost associated with our eating pleasure. At the very least, the price of the foods we import should include the cost of the environmental damage associated with their transport. Such pricing might help us make more environment-friendly decisions about what we eat.

Another significant source of animal protein is seafood. This is a resource which, if properly managed and cared for could provide food for all time without the need for fertilizers, forest destruction, or pesticides. The best practice for harvesting fish is called the "Optimum Sustainable Yield" approach. In this approach, the fishery yields the most fish for a reasonable effort to catch those fish. "Maximum Sustainable Yield" is the number of fish that can be harvested with more time, fuel, and effort, but still leave enough fish alive to propagate for future generations to use, even though it is not the most efficient way to use the resources required to catch them. "Unsustainable Yield" is what happens when we take more of a natural resource than can be replaced, until the resource collapses and we move on to exploit another fish stock to extinction. Unfortunately, there is more unsustainable exploitation than optimum sustainable exploitation in the world today, with fishery resources being depleted around the world. I still recall overhearing a conversation in the days of ICNAF (the International

Convention for the Northwest Atlantic Fisheries Commission) when countries had had a difficult year catching as many fish as allowed under their quotas, so they proposed compensating for the current year's losses by increasing the quotas for the next year. The inevitable result: the North Atlantic cod fishery continued to decline. In northern Pacific and Atlantic waters, some species of fish have dropped by more than ninety per cent since the 1950s.

Another illustration of our "wasted by design" behaviour in this sector of human activity, is the throwing out of food that we consider to be past its prime. Restaurants that have bought too much of something—the inventory exceeds the demand—simply throw it out. Grocery stores that conclude that their inventory no longer looks appealing do the same. Households fail to plan for anticipated consumption without waste, and wind up disposing of tonnes of wasted food each year. Globally, we throw away about 1.3 billion tons of food each year. In America, somewhere between 30% and 40% of food is thrown out each year. This is often deliberate and absolute waste. On the positive side, there are organizations that collect and redistribute food discarded by grocery stores and restaurants and direct this food to proper uses for poor and marginalized people, but this activity addresses only a small fraction of the problem.

Canadians and Americans waste about 400 kilograms of food per person each year. (415 kilograms/person in the US and 396 kilograms/person in Canada.) That's 15 billion kilograms of food wasted in Canada each year and almost 137 billion kilograms of food in the US It is estimated that two-thirds of "bagged salads" are thrown out; half by the stores that sell them because they are stale-dated and the other half by the consumer.

FOOD PACKAGING

One aspect of environmental damage often overlooked is the overpackaging of most food products. The wrapping of individual cucumbers, plastic boxes for tomatoes, labels on individual pieces of fruit, and the wrapping, re-wrapping or wrapping again of meat products, are all examples of deliberate waste in the food industry. For someone who wishes to have a bowl of oatmeal for breakfast the options are to spoon as much as is needed out of a bag of oatmeal or to use a box containing a few "individual servings" of oatmeal. These methods of sale may be convenient or profitable for the producer or marketer, but the environmental costs are borne by others. The problem is that the producer, marketer, or retailer doesn't pay the financial or environmental cost of dealing with the waste; the end consumer pays the cost. The advertiser might be better able to entice buyers to buy something in a shiny plastic package, but he never pays the environmental cost of such packaging. It was not so long ago that a bushel of apples might arrive at the grocery store, to be stacked neatly for the consumer to take only as many as needed; that option is fast disappearing.

The fast-food phenomenon is another particularly troubling environmental issue. More than one-third of Americans eat fast foods each day. These are foods produced and packaged on an industrial scale and they tend to be higher in fats, salt, sugar, and calories than other foods. Of the 300 billion pounds of food wasted in food production, transportation, processing, and consumption processes in the US, 32 billion pounds of waste are associated with fast food outlets alone. Food packaging waste is another formidable environmental problem, with an outsized proportion of such waste, relative to the nutritive value of the food it envelopes, coming from the fast-food industry. Much of the litter on land, in lakes, streams, and rivers, and especially in the oceans, comes from food and beverage packaging, with the fast-food industry responsible for an outsized portion of this.

One of the most wasteful of the packaging problems is beverage packaging in general, and water bottles in particular. Almost all communities in the developed world have spent large amounts of money to provide healthy and tasty water for their residents. Despite this, people continue to buy water in bottles; water that may be less healthy, and that will leave more plastic to be either recycled or, more likely, just discarded. This makes no sense from either an economic or environmental perspective. From the economic perspective, the provision of clean, healthy drinking water would cost less than fifty cents a year per person in the United States. Drinking the same amount of bottled water costs about $1,400. That's an awful waste of money and the environment.

From the environmental perspective, most water bottles in the United States are thrown away, not recycled. Of the 50 billion such bottles used annually in the US, 38 billion become trash. Not only is that waste terribly damaging to the environment, but it also represents the consumption of 17 million barrels of oil to make the plastic bottles, and all the environmental damage associated with that oil extraction and processing. China wastes about the same number of bottles, even though its use, per person, is only about one-quarter of that in the US. World-wide, we use about a million water bottles a minute. How much of this junk can our ocean's take?

Jim Kingham

DEFORESTATION FOR FOOD PRODUCTION

On land, the environment is wounded by deforestation for conversion to food production land. This is not a new phenomenon. The Chinese began doing this centuries ago. Europeans followed suit, cutting down half their oak, beech, and birch trees, as did the Americans. Even the American Indians of the northeast of what is now the United States burned swaths of land to allow the formation of new meadows, which were more useful for increasing the deer population, thus improving the quality of food supplies.

Those activities were generally not disruptive of the dynamic equilibrium that existed between humans and the environment in which they found themselves. The same cannot be said of current deforestation of the Amazonian rainforest. The production of greenhouse gases arising from present deforestation in Brazil is more than the combined contribution of such gases from all the world's ships, cars, trains, and planes. Tropical rainforests have seen more than 60% of their area lost to clearing over the last six decades.

This abuse of the land will not improve. Most experts in the field believe that global population will grow from 7.7 billion people today to around 10 billion in 2050. That's another 2.3 billion bodies to feed. There were 2.3 billion people on earth in 1950. Think of the environmental damage associated with adding another 1950's world full of people to the earth's population in just the next three decades. Most of the new bodies and many of the existing ones will expect to be better fed than is the case in developing countries now. Food production will need to increase by some 70% to meet this need. If we continue along the path we are now on, the area of land that will have to be cleared to do this is equal to all the agricultural land now in use in Argentina, Brazil, and the United States combined. In the 1960s, there were about three acres of agricultural land for each person on earth, now there is only about 1.2 acres of farmland per person, and by 2050 it will probably be around 0.03 acres, half the size of a tennis court.

One area of food waste seldom considered is the growing of food for transformation into fuel for vehicles that are larger than they need to be, driven more often than necessary. One quarter of corn production in the United States is directed to the production of ethanol to help fuel excessive driving in that country. Somehow the use of biofuels like ethanol derived from corn is marketed as being environmentally benign. Looked at in the broader context of carbon use, it is not benign at all.

In light of growing populations and diminishing agricultural lands, powerful countries have tried to buy or lease land in other countries for lengthy time periods (the

Saudi Binladin Group's attempt to get land in Indonesia for rice cultivation, and China's investment in Congo for massive Oil-Palm cultivation, for example).

Just because we *can* consume more calories than needed for a healthy, active lifestyle, doesn't mean we *should*.

Shelter

I use the term "shelter" rather broadly here. In addition to our homes, offices, and commercial buildings that protect our bodies from the cold, rain, sleet and heat, the clothes we wear are included here. I'll start with the shelter closest to our bodies and progress to those more remote.

The *function* of clothes is to afford privacy and to protect bodies from excessive heat loss or gain and the chilling effects of rain. The *form* of clothing is to make us look as attractive as possible. The style and form of clothing is most damaging to the environment.

Clothes are seldom made of animal skins these days, leather apparel and fur coats being the exceptions. Some clothes are made from wool and cotton, but most are now made from synthetic textiles. All paths to the clothes on our backs have environmental implications: the cattle from which leather is produced use up grazing grounds and produce greenhouse gases; the same can be said for animals from whom fur and wool are removed; fertilizers, pesticides, water, and machines are needed for cotton production; and synthetic textiles are made from oil. The apparel industry is one of the bigger polluters on the planet. Textile mills use a tremendous variety of chemicals, many of them persistent and toxic. They are believed to be responsible for about one-fifth of the world's industrial water pollution and use 20,000 chemicals, many of them carcinogenic, to make clothes. Clothing manufacture uses a lot of energy. Since a large portion of the world's clothing comes from China, and China uses coal for much of its energy, there are about three billion tonnes of soot in the air from the Chinese clothing industry alone. Much of the rest of the world's "fast-fashion" is produced in developing countries with little or no environmental protection measures in place. All told, the clothing and textile industry is the second largest polluter after the oil industry. Other human activities, like providing security, are second in line for overall environmental damage as it relates to pollution along with the loss of resources, but the clothing and textile sector comes in second in the *pollution* category alone.

Consumers in Europe, America, China, and Japan buy great quantities of clothing each year to "stay in style." In the production of that clothing, the 15% of material that winds up on the factory floor as trimming waste is thrown away. Consumers throw away clothing at a rate of about seventy pounds per person every year. Thus, the average household in the above-mentioned areas discards almost two hundred pounds of clothing a year, a small percentage of which is because the clothing has worn out. Most is thrown out so people can stay in fashion with the latest designs.

Part of the great throw-away plague is in the realm of young people's footwear, of all things. It was not so long ago that young people in Europe and North America could expect to receive hand-me-down clothing, including footwear. The rational was simple; the apparel was more durable than the wear-out effects while it still fit the child. Today, children often despise hand-me-downs. They insist on having the latest design in many things, including athletic footwear. They are succumbing to peer pressure, driven by relentless advertisements aimed at them by the footwear industry. Here, as in so many instances of environmental damage, the industry is interested only in sales and profit, and bears no cost for the environmental damage associated with the production and ultimate disposal of its products.

Because clothing, especially the fast-fashion sort, is designed to be replaced early because it is out of style rather than worn out, the clothing aspect of personal shelter is a classic example of something that is "wasted by design."

The next level of shelter is that of the family home. There are many opportunities for wasting precious resources and damaging the environment in the design, construction, and operation of the family home. I will not dwell on the environmental effects of lack of insulation, poor quality windows, overheated homes, and so on; more than enough has been written on this subject elsewhere.

A few things are overlooked in the area, however. The trend away from wood-frame structures and the use of more steel and concrete imposes a penalty on the environment because of the considerable amount of energy used in the creation of such materials compared to natural alternatives. Cement-making is responsible for about 6% of the world's greenhouse gas production. Another 8% of those gases come from the steel industry, with half the steel produced going into the construction of buildings. Over its lifetime, about half of the greenhouse gases associated with the building come from its construction, with the other half coming from its subsequent operation. There are ways to reduce such pollution. For example, a window frame made of fir or pine provides four hundred times better insulation value than a plain steel one and is more than a thousand times better than an aluminium one in terms of insulation value.

Similar considerations can be brought to bear with respect to the construction of multi-family, commercial, and industrial units. With respect to multi-family housing, densely packed apartments and condominiums appear to be more environmentally friendly. While this may be so with respect to the footprint their shelter places on the land, it has the negative aspect of putting them in the position of having to rely on outside sources for all their food needs and removes them almost completely from the natural environment. There are places in today's world where one could be born, raised, educated, work, and retire without going into the natural environment. Removed from the natural environment, there is a diminished concern about the health of that environment.

Perhaps the most environment-friendly living accommodation is row housing, two or three stories high, with the minimum exterior wall surface for the living volume enclosed, and a substantial backyard garden space. Street spacing and heat loss from the living space could be minimized while allowing for the possibility of a vegetable garden and a few trees in the back yard. Working and playing in the natural environment, residents could develop a greater sense of the need to protect that environment.

The size of modern homes is an environmental issue as well. The rich and super-rich often like to own "mega-mansions" with tens of thousands of square feet of floor space, scores of bedrooms and bathrooms. Part of the reason for such ownership is to be able to brag to the world how successful they have been in taking a larger share of global resources than their neighbours for both the construction and operation of their mansions. Seldom, if ever, is the environmental impact of building, heating, cooling, and landscaping such mansions considered.

Another environmentally wasteful aspect in the shelter area is the practice of a family owning several homes. Sometimes such ownership is for vacation purposes, as is the case for summer cottages and second homes in distant places with different climates. Sometimes multiple home ownership is for speculative purposes or for storage of capital in a relatively secure jurisdiction. Here again, the environmental destruction associated with the building, heating, cooling, and landscaping of properties that are unoccupied for a large portion of the year is significant but ignored.

The design of larger shelter structures like schools, office towers, and hospitals seldom consider alternative ways of operating such structures in favour of the environment. Buildings are often sealed tight and then cooled in the spring, summer, and fall with large power-hungry chillers and forced-air distribution systems. Providing more openable windows could satisfy the need for temperature control for a large portion of the year. Sealing buildings to isolate occupants from deteriorating exterior air quality is an accommodation which deliberately avoids dealing with the root problem—cleaning up the exterior air. This is one of many examples where we choose to respond to environmental degradation by removing ourselves from the natural environment; in this case, by working in hermetically sealed structures.

Factories are seldom designed with protection of the environment as a primary concern. The driving imperative is to maximize the efficiency and profitability of the factory while passing the costs associated with its environmental damage to the broader community. Hospitals and patients will have to deal with disease arising from factory emissions. Municipalities will have to cope with effluent treatment, or just leave it to cause further health and other environmental effects somewhere down the road. Regional governments will have to pay to size dumps adequate to handle industrial refuse, and so on.

As people design things that relate to our shelter, from the largest office complexes to the most personal "fast fashions," they are almost always ignorant of the fact that they are wasting by design.

Security

Surprisingly "security" is probably the second-worst sector of human activity in terms of damage done to the environment. In addition to the pollution component associated with providing security, there is the environmental toll associated with the extraction of resources and the destruction of land and dwellings in the name of "security." This may seem surprising because we tend to think of our security in terms of personal security and the security of our closest loved ones. Such thinking is terribly outdated in the modern world.

It was the case, thousands of years ago, that we kept ourselves safe by fending off predators and securing lodging that could keep dangerous creatures away while we slept. The development of crude weapons such as spears, hatchets, and bows and arrows made of readily available materials posed little threat to the environment.

With the beginning of the agricultural era, things changed. About 11,000 years ago, in areas just east of the Mediterranean, our ancestors began cultivating crops. The same sort of thing began in the central river basins in what is now China around 9,000 years ago. This tied some people to the land, which provided their sustenance. Buildings were needed to store and protect tools. Granaries were needed to store crops and preserve enough seed for subsequent years.

WEAPONS

Other people were still nomadic, living a rough life by hunting and gathering where they could. An easy form of gathering for some of the roughest among them was to raid the homes and granaries of the new farmers. In response, the farmers developed more defensible buildings and better weapons with which to defend themselves. Walled cities and fortresses followed soon thereafter. The raiders and the sedentary peoples both developed more and better weapons. There was some safety in numbers, so farmers along a given river valley might bond together for collective security, putting their strongest fortifications on the most defensible promontories around their collective farms.

Some security-related activity was not too damaging to the environment: fortifications and weapons were made of natural materials. But building weapons and forts was a waste of materials, energy. and time, when compared with a world in which such human-to-human conflicts did not exist.

The environmental effects associated with community security changed with the beginning of the bronze age, around 5,300 years ago. People began to mine tin and copper, melt these metals separately, mix the melted metals together and pour them into forms or molds to make solid metal objects that were much stronger than things made with either of the starting metals alone. Some of the products made were useful around the home; goblets and pots, for example.

Mining, smelting, and fabrication of metal objects are environmentally damaging activities. We accept the damage, and mitigate it to the extent possible, when the things we make are necessary for the health and efficiency of human activities. Unfortunately, as we evolved into city states, regional entities, and nations, we focused more on resolving differences by force rather than diplomacy. Our ancestors developed weapons with which to most damage their adversaries, seriously damaging the environment as they did so.

A person with a gun in his hands gives no thought to the environment when he pulls the trigger. Someone pulling the trigger to shoot a defenseless animal for sport, or to shoot at the image of another human being for recreation, doesn't think about the environmental impact of shooting. But there is an impact. It begins with the process of gouging the earth to get the minerals from which the weapons and ammunition are made, and the industrial waste associated with their manufacture. Discharge of firearms is a noisy, disruptive activity, whether on the firing range or in the field. The end product of sport shooting of animals or target practice at a firing range is an environment littered with toxic lead. If ever there was something designed to be wasted, it is the bullet.

In the United States alone, there are thought to be some 12 billion bullets made each year. "Billion," remember: 12,000,000,000 bullets a year. That's almost two bullets for every person on earth or more than 37 bullets for every American; every year. That's just bullets made for US citizens—one can only imagine how many bullets per person are made in the rest of the world each year. Enough to kill every human being ten times over, perhaps? Such a waste!

As many as twenty-four different toxic materials are used to make ammunition. Some of the dangerous solvents used in the manufacture of ammunition leak into the environment beyond the factory, while others may be captured for disposal or reuse. Obviously, one of the most abundant and deadly contaminants is lead; lead that injures workers in ammunition production facilities, leaves toxic waste to be disposed of, and leaches into the groundwater after use. Weapons production facilities require careful clean-up procedures to protect the environment when such sites are shut down or abandoned.

This is just the tip of the iceberg. Bullets need guns. Guns need the gouging of countless tons of ore from the earth, belching fumes from steel-making factories—environmental impacts from facilities that manufacture handguns and personal rifles. It is thought that there are about 875 million firearms in the world: 74% of them in civilian hands; 23% in the military, and just 3% in police hands. In the US, there are about 120 firearms per 100 people, four times as many as in Canada and ten times as many as in Russia.

A large part of the reason for so many guns per person in America is the serious wealth inequality there. People at lower levels of wealth see accelerating wealth at the top and want to grab some of that wealth. America has the highest wealth inequality, the top 20% owning 90% of the country's wealth with everybody else owning the 10% left over. Sweden, not far behind in wealth inequity, has only a fifth as many guns per person as in America and spends only one-quarter as much per person on military matters. Part of this discrepancy between wealth inequity and the fascination with weapons is cultural and part is a misinterpretation of the US Constitution, which leads most Americans to think they have a right to bear arms. (The American Constitution actually says: "A *well-regulated Militia, being necessary to the security of a free State,* the right of the people to keep and bear Arms, shall not be infringed." It does not say, "*In order to protect myself and my family I have the right to keep and bear arms.*" Nor does it say, "*In order to preserve wealth inequity*, the right of the people to keep and bear Arms, shall not be infringed.")

In Sweden, personal and national disagreements seem to be addressed in a different fashion. It is illegal for a civilian to carry a firearm, unless for a specific, permitted purpose. Even in those cases where weapons can legally be carried by those with the license to do so, there are strict regulations requiring that the gun be unloaded, hidden, and transported in a safe and secure way under proper supervision. Lethal violence per capita in Sweden is just one-fifth that of the USA. Obviously, the right to carry guns in America hasn't improved security there. Quite the opposite; it has reduced personal security at great environmental (and psychological) cost. In El

Salvador, with an even higher number of guns per person, the homicide rate is many times that of the USA.

While the manufacture, possession, and use of personal weapons reduce individual security and cause great environmental damage, these impacts are but a drop in the bucket when compared with the waste associated with modern approaches to providing national security.

ENVIRONMENTAL COST OF "NATIONAL SECURITY"

"National Security" appears to be threatened by people of lower economic status, different religious beliefs, opposing political philosophies, or those who want to acquire a larger portion of the world's resources for themselves. Instead of allocating time, thought, international coordination, and diplomatic negotiation to address problems, countries damage the global environment by wasting unimaginable resources to advance "national interests" or to defend against the threats of others. The financial costs of defending against perceived threats are many orders of magnitude higher than the costs of dealing with the causes at the root of the threats. The environmental costs are devastating when contrasted with the environmental benefits that would accrue from resolving the root causes.

So, let's look at the environmental impacts of the "security" business. It is indeed a business, at all levels, from the sportsman's gun to the heaviest artillery in the armaments of the most powerful nations. US President Dwight Eisenhower gave us ample warning about this business when he spoke of the Military-Industrial complex almost sixty years ago. Let's start at the bottom and work our way up.

Firearms are small, deadly and environmentally destructive. Artillery pieces are much more damaging to the environment in their fabrication and use and there are tens of thousands of them in the world, each with its own environmental toll during fabrication and with every shot fired. Think of the explosion of gunpowder when fired, and the detonation of explosives when the shell explodes; all at great environmental cost, whether during practice or deadly use. If used in battle, the environmental cost associated with the deliberate destruction at the target end is orders of magnitude larger. Add in all the military transport vehicles, tanks, and other motorized devices to support the military, and we have a "security" environmental impact, from first construction through to daily exercise and operation, that has become quite significant.

The bulk of security related environmental waste is seen in the massive fleets of naval vessels, fighter planes, rockets, medium range missiles, cruise missiles, long-range inter-continental ballistic missiles (ICBM), military spacecraft, army, navy and air force personnel, and so on. Altogether, these represent a total annual world-wide waste of about $2 *trillion*.

Unfortunately, all this military spending doesn't produce security; it destroys security in the long-term in a meta-stable world of mutually-assured destruction. With a nuclear balance here, and an ICBM balance there, rogue states slip in under-the-radar and develop more destabilizing weapons. It also destroys "security" in a more fundamental way, as it destroys the environment.

More than three thousand years ago, two great empires were locked in another environment-destroying battle; small by today's standards but deadly and devastating to the protagonists; Egypt and the Hittites battled for control of the eastern Mediterranean. They negotiated a peace treaty referred to as the Treaty of Kadesk. Recorded in stone in both countries, that treaty survived for eighty years. This was the beginning of international cooperation to end, resolve, and prevent conflict through diplomacy, negotiation, and compromise.

What has gone wrong in the last three thousand years?

Well, among other things, military spending creates great profits for a few people. Politicians persuade their people that such spending is a necessary component of the economy and creates jobs. They almost never speak of the incalculable environmental damage associated with such deliberately wasteful human activity. If the same resources were dedicated to improving infrastructure, alternative forms of transportation, environmentally benign energy production, and so on, much greater economic benefits would accrue, and more long-term jobs would be secured as the environment continues to sustain us.

Surprisingly, one politician of considerable stature recognized where the world was headed sixty years ago. The President of the United States, Dwight Eisenhower, a former army General, no less, warned of the threat to democracy posed by the military-industrial complex. President Eisenhower said:

> Down the long lane of the history yet to be written, America knows that this world of ours, ever growing smaller, must avoid becoming a community of dreadful fear and hate, and be, instead, a proud confederation of mutual trust and respect.
>
> Such a confederation must be one of equals. The weakest must come to the conference table with the same confidence as do we, protected as we are by our moral, economic, and military strength. That table,

> though scarred by many past frustrations, cannot be abandoned for the certain agony of the battlefield.
>
> Disarmament, with mutual honor and confidence, is a continuing imperative. Together we must learn how to compose differences, not with arms, but with intellect and decent purpose. Because this need is so sharp and apparent, I confess that I lay down my official responsibilities in this field with a definite sense of disappointment. As one who has witnessed the horror and the lingering sadness of war—as one who knows that another war could utterly destroy this civilization which has been so slowly and painfully built over thousands of years—I wish I could say tonight that a lasting peace is in sight.

Eisenhower's experience showed him both the destructive soul-destroying nature of combat as a means of resolving differences, and the tragic diversion of resources from world-building and world-saving activities, like building schools and hospitals. As a respected Five-Star General and generally well-liked President, he had seen the potential effects on both sides.

But the economic and lobbying power of the military-industrial complex prevailed (almost 1,000 lobbyists spending over $100 million a year on lobbying) and now the United States leads the way in global military spending, burning up more resources than the combined expenditures of the seven next biggest spenders combined (Russia, China, Saudi Arabia, India, France, the United Kingdom, and France). Thus, the United States, over the last twenty years alone, spent almost $6 trillion dollars, and caused all the environmental damage associated with the wars in the Middle East in the interest of "national security," while spending almost nothing, comparatively, to resolve the issues diplomatically.

Some 2,700 years, ago a wise Chinese philosopher, Sun Tzu, is said to have observed that "the highest excellence is never having to fight, because the commencement of a battle signifies a political failure."

Is it possible to overcome the power of the military-industrial complex of which Eisenhower spoke? Sixty years ago we battled the 'cigarette-industrial complex', which worked so hard to repudiate the evidence brought by the US Surgeon General that "cigarette smoking is a health hazard of sufficient importance in the United States to warrant appropriate remedial action," and won that battle, somewhat. Fifty years ago, the world collectively decided to repudiate the 'refrigerant-industrial complex' which tried to preserve profits at the expense of the environment by denigrating the research of Dr. Rowland and others so that it could continue to destroy the ozone layer with chlorofluorocarbons. Now we are in a battle with the "carbon-industrial complex" which has tried so hard to preserve profits by belittling and denying the evidence that burning carbon is warming the climate. The toughest battle lies ahead, the battle against the military-industrial complex which would have us believe that the only way to resolve international disputes is by maintaining powerful military forces. Imagine a world where all those resources were diverted to keeping the planet's environment livable.

So, when you see your country's artillery fired, see the fleets of naval vessels underway, the squadrons of fighter planes in the air, or the rockets being launched, don't be proud of your country's military power—be ashamed of its wanton destruction of the environment. The manufacture, maintenance, and mobilization of weapons as a means of providing "security" represents the most horrible example of wasting our environment by design. When will we be wise enough to, as Eisenhower put it: learn how to compose differences, not with arms, but with intellect and decent purpose."

Health

There are two sides to the Health and the Environment coin. On the one side, a damaged environment causes health problems. This side of the coin is well-known and often talked about. More than six million people die prematurely each year because of air pollution. About 1.6 million people die in China (360,000 from coal-burning pollution alone) and another 1.4 million in India. In the developed world, every dollar spent on control of air pollution yields between $4 and $30 in savings in health care.

In their sixth *Global Environment Outlook (GEO-6),* focusing on the theme "Healthy Planet, Healthy People," the United Nations draws attention to these unnecessary premature air pollution deaths and the fact that almost 2½ billion people lack access to safe sanitation and 1.4 million die annually from preventable diseases such as diarrhoea and intestinal parasites associated with pathogen-polluted drinking water and inadequate sanitation. Just five years ago there were a billion "open defecation sites" in the world, 600 million in India alone, where people just poop in city gutters or alongside the road in the countryside. Other health effects of our abuse of the environment are often not properly accounted for or ignored. Antimicrobial-resistant bacteria are beginning to be found in drinking water worldwide because antibiotics get into the water cycle through domestic sewage, industrial waste, and agricultural practices. Bacteria mutate to survive in an environment of over-used and improperly disposed antibiotics. Hormone-disrupting chemicals in water affect male fertility and the development of the fetus. There are undoubtedly more effects about which we are unaware.

The flip side of the coin is that our neglect of proper health care leads to damage to the environment. Little attention is paid to this side of the issue.

The health category is in the second tier of environmental sectors important for the survival of the species, after food, shelter, and security, without which we cannot survive to reproductive age. One can survive and reproduce without being healthy. Indeed, that is the condition of many people in both the developing and developed world. On the one, hand we find sedentary, overweight, diabetic, and lethargic young people in rich economies who are quite able to reproduce, and enjoy trying to do so. On the other hand, there are millions of others in the poorer parts of the world, undernourished, suffering from malaria, leprosy, and other treatable diseases, but lacking access to reasonable medical services, who enjoy sexual intercourse because they have little else to spice up their lives.

So, let's begin with the unhealthy diet issue. Every time someone eats more than is necessary to maintain a healthy weight, there is an environmental impact. The demands of agriculture for precious land and water, the impact of agriculture because of pesticide and fertilizer use, the environmental impact associated with transporting, packaging, marketing, and food waste disposal is well-characterised and much of this is described in an earlier chapter. So, every time we damage our health by eating more than necessary, we damage our environment.

Because we eat more than necessary, we succumb to diseases associated with overeating and then require medical attention to treat these diseases. Every such treatment has its own environmental burden. Machines are made that otherwise would not be needed, chemicals are produced to treat things like diabetes and heart disease, and so on. Human resources that might otherwise be directed to saving the planet are directed to saving people who refuse to save themselves.

Humans would be more fit if they chose healthier lifestyles, and the environment would benefit as a result. Instead of driving to the corner store, one might walk or cycle. Instead of sitting in front of the keyboard or television, get out in the fresh air for recreation. Choosing to ride a snowmobile or all-terrain vehicle rather than cross-country skiing or hiking is a negative environmental choice. If one wants a "healthy tan," it can be obtained by gradual and careful exposure to natural sunlight. But too many people choose to use tanning beds instead—teens who use them have a 75% increase in melanoma.

In rich economies, too, many resources are directed to profit-oriented measures to treat disease and too few to prevention activities. Preventing sickness does not generate nearly the revenue stream that treating it does. A comprehensive health care system would devote more resources to encouraging healthier lifestyle choices, which leads to healthier bodies and a reduced dependence on antibacterial compounds, antibiotics, and other drugs. One of the more troubling health/environment interactions is the overuse of antibiotics. We know that to keep ourselves healthy, we need to keep our immediate environment clean and destroy harmful organisms that gain entry to our bodies. Unfortunately, too many people overuse antibiotics as a simplistic solution to problems that should be dealt with more sensibly.

The human body has an innate ability to overcome most organic invaders. Current behaviour denies it the opportunity to develop that resilience. We have overreacted. Despite our ability to overcome most simple infections, many parents put children on antibiotics at the first sign of distress, denying their bodies the opportunity to develop their own defense mechanisms. Some doctors prescribe antibiotics "just in case." Over a lifetime, this creates two problems: the body's immune system does not develop as robustly as it might otherwise, and, for their survival, bacteria mutate to overcome the simple chemical adversaries used to treat infections. I am not suggesting here that parents ignore infections in their children, nor am I offering medical advice. Rather, I suggest that parents take a cautious monitoring approach, in concert with their doctors to see whether a given an infection really requires chemical intervention.

A related problem is that sometimes patients who are legitimately on antibiotics begin to feel better, don't like the side effects of the antibiotics and discontinue them before they have fully destroyed the bacterium for which they were prescribed. The few bacteria that survive such partial treatment mutate and survive as antibiotic-resistant bacteria. When they affect another person, they are a bacterial population resistant to the original antibiotic and more drastic remedies then need to be applied. Thus, overuse and improper use of antibiotics is creating a new family of antibiotic-resistant bacteria. More than 700,000 people die each year from illness associated with bacteria that have become resistant to antibiotics. In the United States alone, 20,000 people die each year from Methicillin-resistant Staphylococcus aureus (MSRA), the best known of the superbugs. We have reached a stage where patients in hospital for routine procedures are at significant risk of dying because of superbugs which came from the improper use of antibiotics.

Overuse and misuse of anti-bacterial compounds in the home environment and elsewhere is another huge problem. It began with anti-bacterial compounds for wiping down food preparation surfaces. Where previous generations had just kept their kitchen counters clean with a little soap and diligent work, the modern cook is encouraged to spray a bit of an anti-bacterial compound on the counter and wipe it dry with a disposable paper towel. Once again, those bacterial not killed on contact are the ones left on the counter and paper towel to mutate and reproduce as anti-bacterial-resistant organisms requiring ever stronger chemicals for subsequent treatment.

Preoccupation with super-cleanliness has reached the level of the ridiculous where people are encouraged to spray their floors with compounds purported to kill "99.9% of bacteria" on the floor. Two things come to mind here. First, when you next step on that floor, whether barefoot, in socks, slippers or shoes, you will simply re-infect the floor with a new batch of bacteria. Do you really need to walk on a disinfected floor? Secondly, that 0.1% of bacterial which were not killed by the anti-bacterial compound used, are those bacteria best equipped to mutate and be resistant to the anti-bacterial compound on its next application. The ultimate silliness in this area is spraying toilets with an antibacterial compound that "Kills 99.9% of bacteria" in the toilet. True, a just-used toilet has a residue of bacteria in it. A properly functioning toilet flushes almost all those bacteria down the tube. Sewage treatment facilities destroy the bacterial before treated effluent is released back into the environment. Why would one want an antiseptic toilet? To eat or drink from it? The next use of the toilet will replace the bacteria that were just destroyed. Perhaps the manufacturers of antiseptic toilet sprays would like consumers to spray their toilets after every use. That might be great for their bottom line, but disastrous for the environment. But then, the toilet-spray industry isn't responsible for the wider environment. Over the last 2,000 years we have learned how to manage human waste to keep ourselves healthy. To now attempt to make toilets antiseptic creates the same problems described above with respect to antiseptic floors and adds another layer of environmental damage as we flush loads of antibacterial compounds into the wider environment with all the attendant bacterial mutation problems, but now on a much broader scale. Those on septic tank sewage systems might like to ponder the concept of antiseptic toilets as they try to have their septic tanks function normally.

Our misuse of antibiotics has led to the emergence of antibiotic resistant bacteria such as *c-difficile*. Often the recommended treatment of *c-difficile* is the administration of more antibiotics. An alternative treatment, fecal transplant, appears to be equally effective and eliminates the likelihood of exacerbating the problem by creating strains of bacteria that are even more resistant to antibiotics. Perhaps physicians avoid this option because of the "yuck factor" (patients resistant to the idea of transplantation of the feces of relatives) and prescribe more antibiotic pills instead. While fecal transplant may be the most effective method of dealing with the intestinal flora disruption problem that may have been caused by over-use of antibiotics, it is also the least expensive, costing almost nothing when self-administered (in which cases it seems to be 100% effective). Yet people are allowed to suffer for years with debilitating diarrhea while they are administered one antibiotic after another as doctors try to wipe out the bacterium instead of letting nature out-compete it naturally. But then, taking the more environmentally helpful, healthier approach wouldn't help the drug companies' bottom line. Besides, the bureaucracy within some of the world's most advanced medical departments hasn't been able to categorize such a simple treatment, so medical practitioners can't use it and bill for it when the procedure is administered by a doctor. Not only are we corrupting the bacteriological component of our environment by design, but we are also causing unnecessary human suffering along the way.

We also flush many chemical substances we use to keep ourselves "healthy" into the environment. Not only disinfectants, anti-bacterials and harsh chemicals, but also contraceptive pharmaceuticals and prescribed hormones. It is difficult to quantify the effects on the reproductive success of unintended recipients of these chemicals, but I recall that when I was Chairman of the Great Lakes Water Quality Board in the 1980s, our scientists were already finding links between hormone-disrupting chemicals in the environment, reduced reproductive success, and developmental mutations. Some of these chemicals, called endocrine disruptors, came from plastics. The *endocrine system* is made up of glands that produce hormones to regulate metabolism, growth and development, sexual function, and reproduction, among other things. One of the strongest correlations we found was between reproductive disfunction and the chemical Bisphenol A, (BPA), commonly found in plastic water bottles. While correlation is not necessarily causation, the sketch below is a possible warning sign, as it shows how human sperm counts have dropped by 60% as the accumulation of plastic in the world grew to more than twenty-five times what it was in 1972.

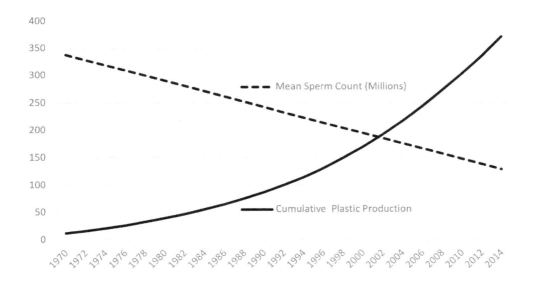

Figure 13. Sperm counts and Plastic Production

Plastics production is hardly the most serious threat here. We can control plastic production, and regulate that BPA be excluded from food packaging and water bottles. The more difficult challenge is preventing the widespread dissemination of hormones and hormone disruptors and their resulting increases in the environment because of overuse of these chemicals in agriculture and in human health programs. More than half the antibiotics consumed are used in agriculture—130,000 tonnes each year. They are used not only to prevent disease in animals crammed into unhealthy, overcrowded cages and pens, but also as a cheap way to fatten up animals for sale.

Even the most modern sewage treatment plants are not designed to remove residues of drugs and other healthcare or cosmetic health products from sewage, so low concentrations of endocrine disruptors are found in the effluents of waste-water treatment plants and in nearby surface waters. It would be very expensive to redesign and operate such treatment plants to completely remove these residues, which affect the reproductive health of humans and many other species.

To the extent that we overuse chemicals we think are necessary for our health and are careless about the release of these chemicals into the environment, we are wasting the environment by design.

Energy

The wealthiest economies in the world have grown fat on a diet of high-carbon energy. Their use and waste of such energy has enabled them to acquire a disproportionate portion of the world's resources for their own use. Many of these countries now resist reallocating a small portion of the global wealth they acquired while damaging the planet's atmosphere to poorer countries so that they might achieve a modest level of well-being without continuing the carbon-rich energy path. Indeed, the wealthiest of the rich countries refuses to even acknowledge that its past carbon consumption has largely created present-day atmospheric problems.

The ability of humans to tap into the chemical energy of fossil fuels has been a boon to human development and the greatest bane for the environment. When humans relied only on the food they ate for all their energy needs, there was no economic growth as we know it. The population was fairly stagnant, and life was extremely difficult. The availability of sufficient energy for human needs has been a critical factor in population growth, the industrial revolution, and economic development.

Over tens of thousands of years, humans have evolved, and their sources of energy have evolved as well. Energy use per person has skyrocketed. The human body typically uses about the same amount of energy as a 100-watt light bulb if human energy use was steadily spread out through the day. That's about how much energy we get from a typical 2,000-calorie/day diet. Since the beginning of time, we have used that energy to think, (our brain uses about 20% of our total energy), to walk and run around (our transportation needs), to search for food and digest it, to build and maintain our shelter, and to reproduce—the most basic needs of any animal.

Sometime in the last 800,000 years, our ancestors, Homo Erectus, started using wood fires for energy to cook and to heat their dwellings. Gathering sufficient firewood was an energy-intensive activity, but the rewards were great, as they expanded their habitats to areas in which they would not have previously survived. If they were able to keep a small fire going, say adding two kilograms of reasonably dry wood to a small fire every hour all day long for warmth and cooking, they would be producing about 8 kilowatts of energy an hour or 192,000 watts of energy per day. For a family of six, for whom energy from their food totalled 14,400 watts per day, the discovery and use of fire increased their daily energy consumption by a factor of more than thirteen. Over millennia, humans learned how to improve their efficiency at gathering wood and began to introduce more reliable and readily accessed energy sources such as peat, and charcoal they created themselves from firewood.

Then, about 5,000 years ago, humans learned how to harness the wind to move themselves across lakes and seas. They also learned to harness horses to carry them over the land. Both of these forms of energy for transportation caused little damage to the environment, although the feeding of horses may have diverted land from other uses. The benefits undoubtedly outweighed the risks.

About 2,500 years ago, Egyptians and others at the Eastern end of the Mediterranean Sea devised water wheels as a means of getting energy "for free" from nature. Once built, the water wheels provided energy for pumping water, grinding grain, and cutting wood for as long as the materials of which they were built lasted. Then they could be rebuilt. Over the intervening centuries, water wheels were improved. The first ones had paddles on the spokes of a wheel with a horizontal axis. The paddles dipped into a flowing stream of water. Later models had the incoming stream of water higher than the outflowing one to provide more power. Then there were overshot wheels where the water came in at the top of the wheel and its weight turned the wheel until the water fell out of the buckets attached to the spokes of the wheels near the bottom of the wheel, giving the maximum possible power from the water wheel. This is the most common type of water wheel seen in museums today.

The final version of water wheels is a turbine enclosed in a water channel at the bottom of a long tube where the incoming water is well above the outflowing water. This is the case where there is a natural waterfall or where a dam has been built to create a water level difference. For more than a century such turbines have delivered a significant portion of the world's electrical power; about 16% at the present time.

Work continues on the refinement of water turbines of different types. Some are now being designed to capture the energy in ocean currents for the production of electricity. For the pleasure of invention, I have been working on a vertical-axis tidal current electrical generator over the last few years.

Windmills began to be used to create usable energy about 3,000 years ago. There was little environmental impact from this method of producing energy for human use.

Drawing from their experience with the use of sails to drive boats over water, our ancestors then moved to having sails attached to the spokes of a wheel mounted on an axle well above the ground so that the force of the wind could rotate the wheel with considerable power. Such windmills provided energy for pumping water, grinding grain, and cutting wood for "free," in terms of the cost of the incoming wind power. The only costs were those of maintenance and reconstruction when the original materials wore out.

The current version of wind turbines is the version seen around the world in land areas where high winds are normal and at sea where conditions are favourable. Wind power now produces about 4% of the world's electricity. The largest land-based wind power facility is in China, designed to produce 20,000 Megawatts of electricity. A Megawatt is a million watts. Using the old "enough energy to power a 100-watt light bulb" example, the Chinese plant would power 200 million 100-watt light bulbs. The largest wind power plant at sea is the Walney wind farm off the west coast of England, with a total capacity of a little over 1,000 Megawatts.

Wind power and water power did little to change the basic energy consumption per human being in terms of daily energy use. One could argue that moving ourselves over water in sailboats, rather than human rowing or paddling, had actually reduced our daily energy consumption.

Of all these ingenious human-developed energy sources, only the combustion of carbon was seriously polluting. Our ancestors knew they were damaging their immediate environment, but the source of this pollution was very useful to them and aided in their development. When they choked on smoke in their cave and huts, they could go out in the fresh air.

With evolving intelligence and fabrication skills, humans could have continued refining and improving the use of natural and renewable resources. Instead, they entered a very black period, both figuratively and literally, for the environment. It was easy to get even more energy from coal and oil (black gold, as it was called in the early days). Further refinements of devices and systems for getting energy from renewable sources ceased. Thus, most grievously, we embarked on a journey to deliberately waste our environment by design, as we designed novel ways to burn carbon instead of using renewable resources.

Since discovering this easy source of energy, we have probably burned about a trillion tons of coal. How much is a trillion tons of coal? Let's imagine we are standing on the top of one of the highest peaks in Europe, the Matterhorn, almost three miles high, on the northern boundary of Italy, looking to the north over Switzerland. Let's imagine that the mountains and valleys of Switzerland are all gone and the entire area of the country, more than 41 thousand square kilometers (km), is flat. The country stretches out ahead of you, 175 km. Looking to the left, the western boundary would be about 140 km away and to the right, the eastern boundary is 240 km away. That's a huge space. Looking down on that flat surface, you see someone digging a hole twenty-one meters deep (the height of a seven-storey building, for those who want an idea of how deep that is. That's a lot of dirt. Now, imagine that the digger keeps going and digs up the entire surface of Switzerland down to a depth of twenty-one meters right across the country from south to north, from east to west. From your vantage point on top of the Matterhorn you see Switzerland as an immense black hole; a hole equivalent in size to a trillion tons of coal. All that coal has already been burned, putting almost three trillion tons of carbon dioxide into the air.

It gets worse.

Not long after exploitation of carbon in coal began, our ancestors discovered how to exploit another source of ancient stored carbon—oil. We have now consumed about 139 billion cubic meters of oil. If we add that to the hole that used to be Switzerland, that makes it more than three meters deeper, and it doubles the volume of CO_2 discharged into the air. Add in natural gas and you would find that humans have dug up or pumped up a volume of material equivalent to a hole across all of Switzerland more than thirty meters deep, (deeper than the height of a ten-story building) with all the associated CO_2 thrown into the air.

That carbon had been drawn from the atmosphere and stored by nature over *billions* of years, but it was dug out of the ground and released by man in just *a couple of hundred* years. It is not surprising, therefore, that the rapid discharge of all that carbon into the atmosphere has influenced global temperatures.

Fossil fuels are hydrocarbon deposits, such as coal, oil, or natural gas, derived from the accumulated remains of ancient plants and animals. They are solar energy, stored as carbon, over millennia. That energy fuelled the industrial revolution, starting around 1760. In the next two and a half centuries, industry and transport poured hundreds of billions of tons of carbon dioxide into the air. Inexorably, the concentration of CO_2 in the atmosphere grew, in lockstep along with the release of that stored energy, and the world warmed.

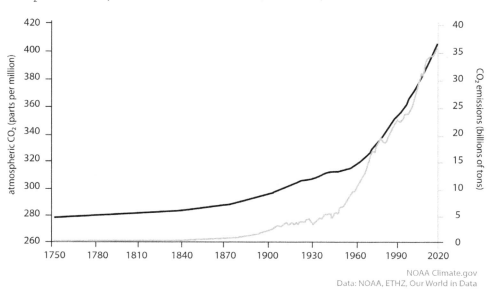

Figure 14. CO_2 in the atmosphere and annual emissions

The burning of hydrocarbons for energy did more than just create CO_2, however; there are a lot of other contaminants associated with that ancient carbon, contaminants that kill humans and other animals much more quickly than global warming. The mining, transportation, and use of coal produces contaminants that lead to sickness and death on a large scale. Particulates, sulphur dioxide, radioactive material, ozone, and heavy metals from coal use alone cause more than a million extra deaths each year around the world. The burning of carbon from oil and natural gas for transportation or heating adds hundreds of thousands more victims to this total.

These risks have been known for a long time. In 1952, London, England suffered from a terrible SMOG event, a combination of smoke (SM) and fog (OG), that made more than 100,000 people sick and killed about 5,000 of them. Most of the pollution in that smog come from burning carbon. Four years later the British Government brought into force a Clean Air Act that made it illegal to burn things that created smoke in many municipal areas.`

Despite the knowledge that SMOG killed people and damaged the wider environment, governments around the world continued to subsidise carbon industries in general, and the coal industry in particular. The rich industrialized countries of the Group of Twenty (the G20) still subsidize the carbon sector to the tune of more than 300 billion dollars a year. A decade ago, the International Energy Agency (IEA) noted that "Fossil fuel subsidies result in an economically inefficient allocation of resources and market distortions, while often failing to meet their intended objectives." Such subsidies are particularly troublesome as they artificially increase demand for carbon resources which damage the environment and hasten the depletion of these resources. These subsidies are hidden from those who pay them—taxpayers. The G20 includes countries like India, China, Turkey, and Argentina, along with other, richer, countries like those in Europe and North America. If taxpayers in those countries were asked to pay an additional tax to deal with global warming, they would resist. In India, for example, it is about $300 per average household per year. Yet they unwittingly pay these sums to ruin the atmosphere because they are paying hidden subsidies to the fossil fuels industry.

If those fossil fuel subsidies were redirected to the development of innovative technologies like more effective electricity storage or artificial photosynthesis there could be significant benefits for the environment. Renewable energy initiatives have been supported at only a fraction of the level of fossil fuel support, notwithstanding the environmental benefits they confer. This is yet another example of how mankind is deliberately supporting projects that damage the earth's environment and its storehouse of non-renewable resources. Yet another clear example of the environment being wasted by design.

The emergence of cheap, but damaging, energy from ancient carbon deposits deflected our attention away from the former wind and water energy sources and ways in which such energy could be stored and used more efficiently. One of the reasons it was so cheap is that no one thought about the cost of disposing of the waste products of carbon use; that is, the disposal of carbon dioxide into the atmosphere. Early in the twentieth century, about 35% of cars in America used electrical energy, stored in batteries in those cars. Cheaper internal combustion engines drove that number down to zero because the costs of dealing with the emissions from gas and diesel engines were not paid for by the producers or drivers of such vehicles—they were just passed on to the next generation.

Had the trillions of dollars poured into the exploitation of fossil fuels by industries and governments over the past century been directed towards refinements in the production, storage, and use of energy from renewable resources, the electric car might have prospered as the preferred form of personal transport over the last 120 years. But, as usual, our predecessors chose the laziest and cheapest option, fossil fuels, without a thought to the environment. Thus, mankind became addicted to fossil fuel energy. That addiction, and its damage to the atmosphere, continues.

Carbon Emissions by Country
(Megatons CO_2 in 2017)

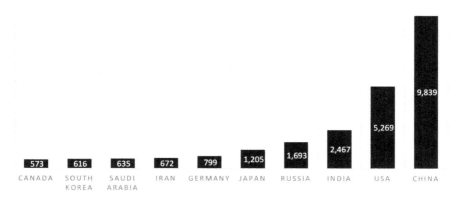

Figure 15. Carbon emissions by country

In 2017, Canada pumped 573,000,000 metric tonnes of CO_2 into the air, about one-ninth that of the USA, or about one-seventeenth that of China. But this is not the right way to look at how damaging individual Canadians, Americans, or Chinese are when it comes to their contribution to global warming. The picture changes dramatically when one looks at the amount of CO_2 discharged by an average person in each of those countries.

Per Capita CO_2 Emissions (Megatonnes -2018)

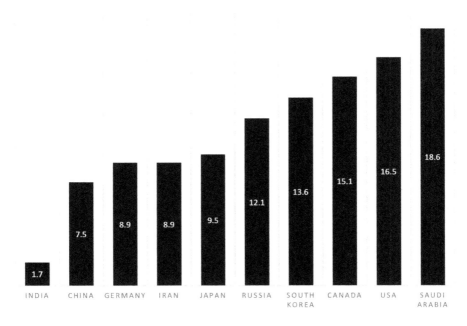

Figure 16. Per Capita CO_2 emissions

It is apparent from this chart that the average Chinese person is responsible for less than half the CO_2 of the average American. When Americans argue that they should not have to pay to help solve the world's CO_2 problem because China is the biggest polluter and should therefore pay more than they do, they are arguing that the average person in China, who is adding only half the CO_2 of the average American, should have to pay more to put an even smaller proportion of CO_2 per person into the atmosphere. For China to equal America in terms of CO_2 emissions, the average Chinese citizen would be allowed to emit only one-quarter the amount of the average American. This does not excuse either country, or other big emitters like Canada, Saudi Arabia, South Korea, and Russia for not taking action to reduce these emissions. Rather, it is a suggestion that those who point fingers at others for their emissions should first examine their own per capita emissions.

THE NUCLEAR POWER ISSUE

In the mid-twentieth century, a more promising alternative to fossil fuels for electricity production emerged. That alternative was nuclear power. In this case, governments saw potential threats to human health and the environment and imposed heavy burdens on the nuclear industry to minimize them. One had to ensure that the release of radioactive substances from nuclear power plants was minimized, and that plans were made for the safe storage of spent fuel from reactors. Nuclear power generates no greenhouse gases as it generates electricity, though there are emissions associated with uranium mining in the same way that there are emissions associated with the mining of carbon as an energy source.

Why has nuclear power not displaced carbon power? Primarily because of a misplaced fear that nuclear power is somehow more dangerous than carbon power. It is true there have been accidents at a few of the many nuclear power plants around the world. These accidents and other aspects of the extraction, use, and disposal of uranium for nuclear power probably caused about 5,000 deaths in the period between 1971 and 2009. That is a big number and a sad thing to contemplate. However, over the same time period, the use of fossil fuels has undoubtedly caused millions more deaths than would have been the case had nuclear power been used instead—far more tragic.

Our choices with respect to energy production stand in stark contrast to our decisions with respect to long distance travel. The two have an interesting parallel. In long distant travel, we know that every now and then there will be a catastrophic air accident resulting in the loss of a few hundred lives. We also know that each year millions of lives are lost in alternative forms of transport. We continue to use both forms, and the number of people travelling by air still increases year by year, even though passengers know that they have no control over situations that might lead to an air crash. It is still safer than travelling by car, train, or ship, per passenger mile travelled. Contrast that with the energy situation in terms of nuclear energy versus carbon burning. Every now and then there is an incident at a nuclear power facility and a few hundred people may be affected. In very rare cases some will die because of a nuclear accident. But we know that every year millions of people will be killed as a result of producing power by burning carbon, especially coal. We have been persuaded by the carbon lobby and the governments they support that the small risk of some people being affected by a nuclear power plant incident justifies continued reliance on a far more dangerous energy source—carbon.

ENVIRONMENT-ENERGY-ECONOMY LINKS

Why, then, do we use the more damaging energy source? Economic considerations are front and centre here as we continue to ignore societal and environmental costs associated with burning carbon. When politicians see emission reduction at odds with economic growth, they always choose economic growth. But this is a false choice conditioned by the political time frame. While it may be the case that controlling emissions may hamper economic growth over the next four years, not controlling emissions will have a far more negative effect on economic growth over the longer term.

In 1997, an agreement to deal with carbon dioxide and five other man-made gases that warm the atmosphere was ratified by 140 countries. The United States was by far the world's leading emitter at the time. Here, too, when the crunch between the environment and the economy surfaced, the American President pulled his country out of the agreement. "Kyoto would have wrecked our economy. I couldn't in good faith have signed Kyoto," President George W. Bush said. Confirming the observation made earlier about our addiction to carbon-sourced energy, he went on to acknowledge "We're hooked on oil…"

Another consideration that impedes the substitution of nuclear energy for fossil fuel energy is a misplaced concern about the disposal of nuclear waste (a political rather than practical problem) and sensationalism whenever there is a nuclear power plant incident (sensationalism undoubtedly amplified by the powerful carbon lobby). These comparatively rare and avoidable accidents seem to trump the clear downside of millions of excess deaths assured through carbon use and the widespread and long-term environmental damage arising from our on-going carbon addiction.

It is likely that the use of fossil fuels for electricity generation will continue for many years. Even though we know how damaging they are to the environment we still rely on them to produce more than 60% of the world's electricity. Coal generating plants alone are responsible for a quarter of the world's carbon dioxide emissions, and those emissions are increasing. With energy consumption running at between 4,000 watts and 25,000 watts per capita in the rich world we are now using 40 to 250 times as much energy per person as did our ancestors before they discovered fire.

We are not only wasting the environment by design in the way we produce energy, but we are also deliberately wasting the environment by wasting much of the energy we use. Some wasteful practices are blatantly obvious, while others are almost impossible to discern. In the first category, think of neon signs and billboards lit up through the night. Advertisers must think that somehow a lonely traveller in the middle of the night may happen to see that sign and say "Hey, I need to buy one of those!" A faint hope to be sure, but as long as the advertiser can get cheap energy, he will take the chance.

At the opposite end of the scale, the "almost impossible to discern" end, we have the misuse of the internet. Every time an internet user views a movie on-line or someone plays an on-line video game, internet server centres are whizzing away. They consume formidable amounts of energy and generate millions of BTUs of waste heat, which requires the consumption of even more energy, as air-conditioning is necessary to keep the server centres cool. "Mining" coins for the crypto-currency, Bitcoin, uses as much energy each year as Austria does. But users are hardly aware of this waste, if they think of it at all. To be sure, producing hard copies or CDs consumes more energy per word. But a great deal of internet energy and bandwidth is used to share on-line videos and games. Every hour spent playing at a computer console is another hour of quality time out in the real environment lost, and another hour's worth of internet server energy wasted. At the current pace of expansion, internet use may well consume one-fifth of the world's energy by 2025 and produce 15% of greenhouse gases. Half of the electricity used in the world's internet server "farms" goes to running the servers themselves; the other half goes to running the air conditioning units that keep them cool. The heat ejected by the air conditioning units goes directly into the atmosphere, adding to global warming.

Between these two extremes are wasteful practices that are part of our everyday life. There are millions of such activities. A simple one is the unthinking use of clothes dryers during peak hours of electricity use, instead of during times when demand is low. If a city of 250,000 people switched to drying clothes during off-peak hours, an entire coal-fired power plant could be shut down. An even less wasteful alternative would be to switch to the old solar and wind powered clothes dryer—the clothesline.

To summarise, then, we produce about 170,000 Terawatt-hours of energy around the world. That is 170,000,000,000,000,000 watts, every hour of every day. It is difficult to even imagine that amount of energy. Even the old metric of: "Enough energy to keep 1,700,000,000,000,000 one-hundred watt light bulbs lit" is too big to imagine. It's the equivalent of keeping hundreds of thousands of Boeing 747 airplanes in the air, every hour of every day. That's a huge amount of energy, almost 90% of it from burning carbon while polluting the air with CO_2.

The next chart shows how we have doubled our energy production and use over the last fifty years. This energy comes from five sources: oil (33%), coal (30%), natural gas (24%), hydroelectric power (7%), and nuclear power (4%), with the remaining 2% from wind and solar power. Oil, coal and natural gas have obvious environmental effects. Hydroelectric power can have damaging effects as well. Nuclear power has more difficult political problems than environmental ones.

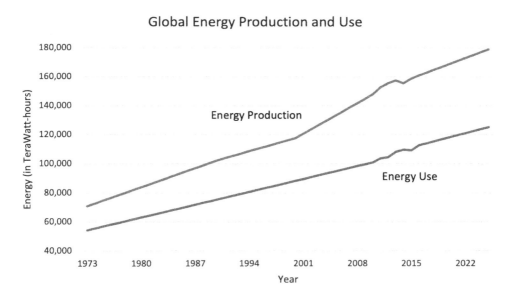

Figure 17. Global energy production and use

It is very difficult, if not impossible, to get reliable values for the way energy is used around the world. Averaging different sources of information suggests that about 42% goes to industry, 28% for transportation, 15% for residences, and the rest to commercial operations. The energy used for transportation is particularly damaging to the environment.

In OECD countries, for every 100 gigawatts of energy produced, the death rate from accidents is less than one for nuclear energy, more than seven for natural gas, almost nine and a half for oil, and twelve deaths per 100 gigawatts of coal-produced energy. In the United States alone, premature deaths attributed to particulates released by fossil-fuelled power plants are estimated at 30,100 per year. Add to this more than 4,000 cases of pneumonia, 9,720 hospital admissions for cardiovascular ills associated with these particulates, 59,000 cases of acute bronchitis, 603,000 asthma attacks, and more than five million lost workdays.

Perhaps these numbers are too abstract to be meaningful. Let's put them another way. Each day in America, on average, more than 14,000 people miss work because of the deleterious health effects related to fossil-fuelled power plants. Along with that is the considerable cost of health care associated with eleven new cases of pneumonia, twenty-seven cases of cardiovascular ills, eighty-two premature deaths, 161 new cases of acute bronchitis, and 1,652 asthma attacks. *Every day.* The World Health Organization did a comprehensive study of the number of deaths per terawatt hour of electricity produced for every major energy source. They concluded that nuclear energy had the safest record of all energy sources, including wind and solar power. Indeed, a paper published in a journal about environmental science and technology concludes that as many as 1.8 million premature deaths have been avoided thanks to nuclear power displacing some of the fossil fuel power over the last fifty years.

"What about nuclear accidents like Japan's Fukushima Daiichi?" you may ask. That incident may have led to a thousand premature deaths, according to some experts—less than one-thirtieth the deliberate annual death rate from burning fossil fuels for power generation in the United States each year. Even this understates the problem. If Japan had generated the same amount of power over forty years from fossil fuel rather than from the Fukushima complex, there would have been scores of fossil-fuel related deaths, along with formidable health care costs associated with all of the other fossil fuel-related illnesses described earlier.

British Petroleum's most recent Statistical Review of World Energy showed that for 2010, energy demand grew by 5.6% across the board. This is the fastest annual growth in almost four decades. And this at a time when countries are saying they intend to reduce carbon emissions well below current levels. When you look at the details, you find that the least carbon damaging of the major energy sources, nuclear energy, grew least, at about 2%; oil and gas grew at about 5% and 7% respectively; and the dirtiest, coal, at almost 8%. Obviously, we have not been serious about reducing carbon inputs to the atmosphere.

There is a solution to the carbon problem—treat it like other waste disposal problems and charge for the amount disposed. Of course, it is a bit tricky when one must deal with the discrepancy of use between the most developed and the least developed nations on earth. It is not fair to charge the same amount per tonne for disposal in a country where emissions are ten times greater per person than in another, poorer, country. The international managerial skills and financial capability to deal with this discrepancy exist, if the collective political will to do so would emerge.

One could take the global average of carbon emissions per capita and charge nations above that average a reasonable amount, say $20 (USD) for every tonne *per capita* over the limit. The funds thus set aside would be held for reallocation to those countries below the limit on the basis of $20 per tonne below that limit. But such funds would not simply be divvied up and given to the recipient countries. Rather, the United Nations Environment Program, UNEP, working in concert with the International Energy Agency, IEA, could make the funds available to those countries for *bona fide* approved carbon reductions in each country. This approach would have simultaneous beneficial effects in both the developed and developing world. The 'developed' world would have a powerful incentive to reduce carbon emissions to reach the mean and developing countries would have incentives to develop in a direction that would be less carbon intensive. To respect the sovereignty of nations, it would be left up to them to decide how costs or benefits would be allocated within their jurisdictions. Some might choose to penalize heavy vehicle use, others to reward efficient building construction.

Such solutions to global energy problems will be challenging to implement because of political issues. Rather than focusing on solutions to these problems, governments around the world have continued to subsidize fossil-fuel production. Perhaps as much as a trillion dollars a year is paid by taxpayers, whether they like it or not.

For most of human history, people lived with no "artificial" energy sources. Life was rough, uncomfortable, and cruel. Tapping into fossil fuels made life much more prosperous for our generation. As we discovered alternative energy sources such as water, wind, and solar energy, we had the option of choosing those which would be best for the environment. Unfortunately, governments driven by the singular goal of increasing GDP above all else, and prodded by those with a vested interest in fossil fuel exploitation, chose the easier, lazier path without regard to the environmental consequences.

Thus, in this area of human endeavour more than any other, the environmental commons has been wasted by design.

Transportation

Transportation is not essential for survival.

Part II of this book is structured to address the effects of our activities on the environment in order of their importance for our survival. Nourishment, shelter, and security are crucial. Good health is important for survival, but not as critical. Artificial sources of energy were not critical for most of human existence, and are still not essential in some remote areas, although most of the world has become dependent on such external energy supplies. Transportation falls into the same category as energy. Transportation was not essential for survival, but we have allowed it to become a critical component in modern life. It accounts for more than a quarter of our energy use and CO_2 discharge.

For a very long time, humans got around by walking and running. That was all that was needed to gather food, chase prey or escape predators. As the water environment began to be exploited for food, our ancestors developed rafts and dugout canoes to improve their food-gathering efficiency. This extension of our personal mobility had very little impact on the environment. Adding sails also aided transportation with the use of natural elements; again, not harmful to the environment. Improving mobility on land began a process of environmental degradation that has continued and accelerated over the last few millennia. The use of horses required the provision of feed, shelter, and the management of manure. Stress on the limited supply of arable land began. But the stress on the environment was still relatively minor. Not so today. We have evolved to require transportation systems that are unnecessarily and extensively damaging to the environment. Modern transportation is one of the most damaging aspects of human activity, with most of that damage resulting from deliberate waste.

A division of labour emerged that favoured gathering large groups of people together to provide goods and services, while leaving most others in the countryside to provide food for all. That generated a need for transportation within and between these groups. Initially, much of that transportation was "public." There were multi-person carriages, streetcars, and trolley buses in cities, and horses and wagons in the countryside. In cities, people tried to live as close to their place of work as possible and many services were located as close to these groups of people as feasible. Primitive railroad systems provided transportation between urban centres and allowed for the transportation of bulk food supplies to the cities.

All these forms of transportation had serious environmental side effects. Some, such as the collection and disposal of horse manure, were recognized and dealt with. Others, like the emissions belching from steam engines, were recognised but ignored. Most of the long-term environmental effects of transportation were not even recognized.

Not so today. Now we recognize, study, understand, and continue to largely ignore the horrible long-term environmental effects associated with our present-day presumption that every individual is entitled to travel anywhere in the world, wherever and whenever he or she wishes. We are now expanding our horizons to offer to the very rich the privilege of travelling to outer space, at an outrageous environmental cost to be borne by all, including the very poor.

Serious environmental problems began with the invention of the internal combustion engine. This marvel of innovation and technological development displaced earlier transportation power sources because of its convenience and effectiveness. The cumulative environmental costs were not considered, and no one paid those costs while they enjoyed the benefits. In one city alone, Delhi, India, the burning of diesel fuel accounts for almost 80% of the most damaging particles in the air. These particles probably cause 300,000 excess deaths each year in Delhi, and two and a half million deaths a year across India.

Transportation now accounts for 28% of the world's energy use.

To this day, people around the world resist the call to pay the environmental costs associated with the dumping of waste gases into the environment. Often, they also resist the suggestion that the price of heavy crude oil from tar sands should include a fee for the complete remediation of the areas from which they are extracted. Instead, billions of taxpayer dollars are spent to subsidize the industries that cause this form of pollution.

Here is where the conflict between the environment and the modern economy comes into sharpest focus. There is money to be made—a great deal of money—in the manufacture, fuelling, servicing, and marketing of automobiles for private use. There is not nearly as much money to be made in the provision of streetcars and buses for mass transportation. The environmental costs of allowing—indeed encouraging—every individual to move his or her body in a two-ton car anywhere and everywhere, at any time, have been deferred to future generations. We all know this. To entice and stimulate the purchase of new vehicles, manufacturers advertise cars that can climb mountains, ford streams, plough through the muck, zip along from zero to a hundred kilometers an hour in mere seconds. Consumers are being conditioned to buy faster, bigger, heavier automobiles (the average car gained 800 pounds between 1987 and 2010) loaded with fancy features like on-board video screens and "surround sound."

What is being sold here? Presumably one buys a vehicle to provide safe, efficient, comfortable, economical, and environmentally friendly transportation for a few people to travel from home to work, shopping, or to visit friends, and so on. So why does one see so many light trucks and oversized SUVs on the roads and clogging up parking lots? Why do advertisements for personal transportation feature monstrous vehicles tearing up sand dunes, forest streams, smashing through snow drifts or climbing mountains? It must be that such ads sell cars. How often do you see advertisements featuring the quietest engines, the best fuel economy, the minimal use of raw materials for construction? Could it be that brute force appeals to the buyer's imagined self-esteem with nary a thought for the resultant environmental damage? Personal transport is designed to waste the environment, rather than provide efficient transportation.

Instead of making public transport more convenient, more effective, and more ubiquitous, limited resources are directed to covering the landscape with roads and parking lots. This imposes a very serious cost to everyone, not just for the roads and parking facilities, but for lost time trying to get places and finding a parking spot. One study of this problem, done by the Texas Transportation Institute, said that in 2007, traffic congestion in the United States cost Americans 4.2 billion working hours and 10.6 billion litres of fuel—such a deliberate waste. Properly designed and run public transit systems would prevent most of that waste and could offer a much more pleasant commuting experience.

The most efficient way of moving a human being from one place to another is by bicycle. Cycling gets people into the environment and is twice as efficient in moving a body over the ground as is walking. Half as many calories burned per kilometer to move someone by bicycle, means half as much CO_2 exhaled by the cyclist. That's one-twentieth the CO_2 produced by moving that same body from one place to another in a car. It is healthier than driving and, for much of the year, can be a pleasant experience. To make it more pleasant, cities can do much to improve the comfort and safety of cyclists. Unfortunately, in North America, the risk of an accident for a cyclist riding to work is four times greater than for cyclists in European countries such as Germany, Denmark, or the Netherlands because most American cities are not bicycle friendly. European cities have bicycle lanes protected from fast moving cars, and bicycle-activated traffic lights at intersections that give priority to cyclists. So, Europe is on track to improve the health of commuters, reduce energy use, and help mend the environment, while most North American cities continue to pander to the automotive and fossil fuel industries and deliberately waste the environment.

Here's a chart showing the relative amounts of CO_2 for different modes of transport.

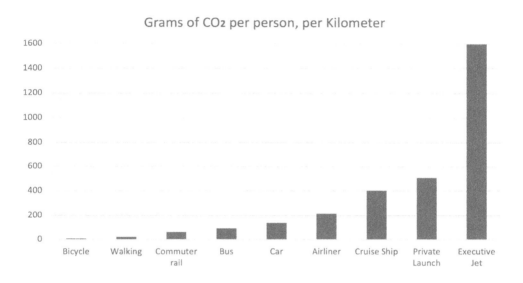

Figure 18. Grams of CO_2 per person, per kilometer

A few words about this chart and the assumptions used in making the calculations: the most damaging form of transport, the executive jet, uses one of the most common of such planes and assumes there are seven people aboard. They can't hold many more than that and when they hold fewer, the environmental damage per person goes up accordingly. Similarly, it is assumed that the 33-foot trawler-type private motor launch has two people on board. The fuel consumption for a 33-ft planing-hull launch at speed is probably three or four times more, per kilometer. For the cruise ship, it is assumed that there are 3,000 people on board, each one of them responsible for 402 grams of CO_2 per passenger mile. It is assumed that the commuting car contains one or two (average 1.5) people, the bus contains twenty people, and the commuter train has two hundred passengers. The airline example is based on 190 people aboard a plane like the Boeing 737-800.

One can make different assumptions and consider different factors. For example, how much energy does it take to make a car in America rather than China? How long does the car last? What is the total lifetime carbon burden per person per kilometer over the car's lifetime of repair and fuel use? Similarly, different assumptions can be made for most of the forms of travel in the chart, but different assumptions do not change the broad picture significantly. If you want to get to the office, driving your own car is not the best choice from an environmental perspective. It is certainly not a great idea to go across the sea in a private jet or motor launch.

One of the more serious deliberate assaults on the environment is the modern fad of travelling anywhere in the world anytime we want to. This is a relatively recent phenomenon. For most of our existence, humans travelled to get food, find shelter, or escape predators. Now we go "for a holiday" to see something new. University students (who should know better) zip off to the Caribbean during the spring "study break" for a wild bash. Businesspeople travel around the world several times a year to meet face-to-face with their colleagues. About ten million people take an airplane trip each day (5.5 billion trips in 2019, more than one trip for every two people on earth). That's double the number from just a couple of decades ago. The energy costs of such travel are formidable. The environmental consequences are disastrous.

Fuel consumption per passenger for air travel is about three liters of aviation fuel per one hundred km; more for short-haul flights, less for medium length flights and more for very long flights where the plane needs to carry much more fuel (which is heavy), and therefore fewer passengers, to complete the flight. Burning this fuel produces carbon dioxide, carbon monoxide, nitrogen oxides, and fuel particulates that damage the environment, as well as water vapour, which affects the energy balance in the atmosphere. Because long-distance airplanes fly near the stratosphere, their damaging effects are amplified. The number of airline passenger trips grew relentlessly until 2020 when the COVID crisis saw it nosedive. By 2019, *tourist* travel alone had reached 1.4 billion trips a year, with 8.5 trillion air miles travelled overall. COVID pushed the numbers back down to 1990 levels. But the world survived without all those air miles travelled. The air environment benefitted. One has to ask: Do we want a ruined environment a century from now because everyone in *this* generation gets to go wherever in the world they want, whenever they want?

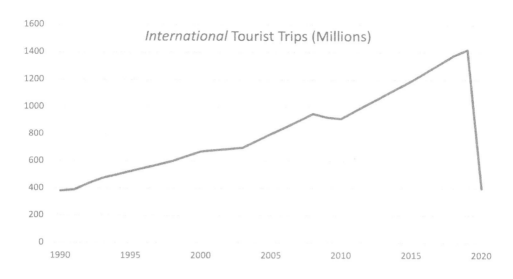

Figure 19. International Tourist trips.

Passengers do not pay anything near the cost of the environmental damage associated with their air travel, even if they participate in some kind of "environmental damage offset" program, like planting trees in the Amazon, for each flight they take. As air travel increases at a steady rate, year after year, the environmental damage grows at an alarming rate. The international community has refused to apply greenhouse gas emission penalties to the airlines, thus keeping passenger costs artificially low.

We know this. Most passengers know this. They also know that there are holiday options closer to home—in their own backyards perhaps—yet they continue to exercise their "right" to travel and leave the poorest of the poor to pick up the environmental tab as homes are flooded and farmlands parched in the poorer parts of the world. Of all the measures to protect the environment, reducing or eliminating long-distance travel is one of the two top items that elicit an almost immediate negative response when you discuss this with friends and colleagues. Their responses: "You're not suggesting *I* shouldn't take my cruise this year!" "*My* not going to Paris this spring isn't going to save the environment!" "Don't try to change *my* holiday plans!" Sadly, as *everyone* says this, the environment is being wasted by design.

There are alternatives for holiday seekers, but if they refuse to recognize the damage they are doing, they do not accept the alternatives. The computerised rich world, with virtual reality presentations of practically every tourist destination, large-screen TVs in most homes, and surround sound, offers details of faraway places more clearly than can be seen from the seat of a crowded tour bus. Businesspeople can now use virtual meeting rooms that are so realistic that it seems that participants across the world are actually present. As long as airlines and tourist destinations continue to profit from tourism, they will continue to promote long-range travel.

Not satisfied with travelling to the other side of the world without regard to the environmental implications of such travel, the richest people in the world are travelling to space—just because they can. It will be risky, it won't be a comfortable trip, but it will be an adventure they can brag about if they survive. The environmental costs, to be borne by others, are almost incalculable. From the construction of the massive vehicles that will thrust them into space, to the tons of emissions per person and the disposal of all the waste left in their wake, both on earth and in space, there are formidable environmental effects.

For moving freight, we rely far more on trucks than need be. The problems here are similar to those associated with commuter traffic. Our roads, parking lots and cargo transfer structures have been made most amenable to truck transport, while rail transport continues to be restricted despite its much higher fuel efficiency per tonne of cargo moved.

We have moved from a relatively non-mobile society to an extremely mobile one. Whether for commuting, shopping, tourism, or shipping freight, transportation has become perhaps the most environmentally abusive of human activities. Here, more than in most other areas, this environmental waste is deliberate—the environment is being wasted by design.

Consumer Goods

We used to call them durables…

I grew up in a rural setting, in a family of very modest means, on the outskirts of an average-size town in Ontario. It was a great life. We may not have been rich, but we didn't feel poor. Because I had an older brother, I learned at a very early stage to appreciate the value of hand-me-down clothing. We were blissfully unaware of designer running shoes and the latest fashion in windbreakers. Most things were second-hand, and that was okay. We didn't have a car until I was twelve, but the bus service was great. In my mid-teens we got a television; we never had more than one. Until then, the radio had provided many an hour of imagination-stimulating programs from "The Shadow" to "The Happy Gang." Grocery stores, hardware stores, and dry-goods stores had all the necessary things for a good life. It seems the old stove and fridge lasted a lifetime. They were called "durables" in those days because, well, they *were* durable—they lasted. Nowadays they are called consumer goods. They are consumed, they break down, they cannot be easily repaired, they quickly wear out and then are discarded.

My grandchildren know only this new world (despite a few of Grandad's comments: "When I was your age…"). While I don't begrudge them the variety of experiences and the extent of information available at their fingertips, I do regret that so much of their material world has come from a huge commercial consumer products system that has unnecessarily and deliberately laid waste to non-renewable resources. One of the routes to the success of the consumer goods business has been to continually introduce "new and improved" models of various products that render their predecessors obsolete. Not only do computer manufacturers practice this technique, but manufacturers of big-ticket items like televisions, major appliances, and automobiles rely on it as well.

Advertisements are becoming ever more effective at identifying potential customers and seducing them to spend money—using troves of data that have been gathered from internet platforms to identify their consumers' wishes.

Part of the deliberate waste arises from the overwhelming variety now presented to entice consumers to purchase goods, whether they need them or not. One can buy a hundred different cell phones or fifty different kinds of television. Scores of brands and styles of running shoes are now available for children to demand, so that they may move with the "in" crowd. Thousands of different makes, models, and colours of automobiles, with their many and varied accessories, entice people to junk used cars and buy something new—just because it is new. In 1975, North American consumers had fewer than 10,000 items to choose from in a typical supermarket, and yet that seemed more than enough to satisfy their needs. Now there are about 50,000 items available. Not only is that an unnecessary number of items, it is a mind-numbing variety that confounds the purchaser. Behavioural scientists have concluded that the myriad choices presented to the modern consumer are counterproductive—they are either unable to decide at what is best, or they just grab whatever comes to hand first.

Barry Swartz has addressed this problem in his book *The Paradox of Choice* in which he observes that "choice no longer liberates, but debilitates. It might even be said to tyrannise." The assumption that some choice is good, therefore more choice must be better, is called into question here.

An overly large number of consumer devices are now powered by batteries, most of which are single-use batteries to be discarded when they run down. To illustrate, in the United States alone, more than fifteen billion batteries are sold each year. About 100 million of these are lead-acid batteries. Think about the environmental toll of mining the necessary lead, fabricating the batteries and then disposing of the waste—both the wasted batteries and all the by-product wastes associated with their manufacture. Of the remaining 14,900,000,000 batteries, three billion are used for portable radios, and most of the rest are for other portable devices, like mobile phones, more than 150 million of which are thrown away each year in America alone. It was not so long ago that one never had to worry about a watch battery wearing out; there were no watch batteries. One did have to remember to wind the watch once a week. Now many people often choose to throw the watch, battery, and strap away when the battery expires because it's too difficult to find the right replacement, too difficult to install, the watch band is worn out, or a new style watch is much "smarter." What was wrong with winding a watch that lasted for years? What was wrong, from the producer's point of view, was that it didn't stimulate demand to buy a new watch every year.

Then there are portable computers and a host of other electronic devices disposed of each year. If you have a good imagination, try to imagine a gigantic weigh scale. In the pan on one side, put the electronic waste discarded each year. To balance the scales, you would have to stack seven of the great pyramids of Giza on the other side. That's an awful lot of deliberate waste.

It is not the mere existence of choice in consumer goods that is troubling from the environmental perspective, but rather the concomitant waste of resources and the associated misuse of limited energy supplies in providing those choices. We have uncritically accepted the mantra that choices between dozens of variations on the same item drives competition and makes markets work, generating economic growth. Too much emphasis has been placed on a market economy driven by the acquisition of material "things" rather than meaningful "experiences." In this, we are markedly different from aboriginal peoples who enjoyed a love of the land on which they lived rather than a love of things that could be made by destroying the land.

Secular holidays like Mother's Day, Father's Day, or Valentine's Day are little more than commercial opportunities to buy more things. Religious holidays are even worse in the sense that we are encouraged to start spending early ("Only 100 shopping days 'til Christmas") and spending heavily. Halloween has become one of America's most popular holidays and spending is rising as more adults partake in the festivities. Spending on Halloween costumes was expected to reach $2.8 billion in 2019, according to the American National Retail Federation, and total spending for the holiday was expected to grow to $7.4 billion, up from $6.9 billion the previous year. Most of these purchases end up in the trash bin.

Another part of the consumer goods problem is the fabrication of goods that cannot be easily repaired, so that the cost of repair exceeds the cost of purchasing a new item. There were many examples of this in the list of items in Chapter 8 "Everywhere you look." There are thousands, if not millions, more such examples. Think about this every time you seek to have an item with a small (probably planned) fault repaired and are told that it will be far more expensive to repair the damaged item than to throw it away and buy a new one. Think about this when you seek to replace rechargeable batteries in a device for which they were uniquely designed and are told that it will be cheaper to buy a new device complete with rechargeable batteries, charger and carrying case, than to buy a replacement set of rechargeable batteries. Think about this when you see a that small part of an assembly in a costly item is broken and are told that it cannot be replaced as a single part—the entire assembly will have to be replaced. Think about this the next time you see a garbage truck on its way to the dump loaded with heaps of stuff that had to be thrown away because it couldn't be readily repaired.

This is also apparent in the large appliance field. Things like refrigerators, stoves, washers and dryers, wall ovens, and so on. Landfill sites have a special area with acres of discarded large appliances, many of them just off-warranty, because they are too difficult or costly to repair, or because the manufacturer does not sell the unique replacement parts needed for repair, just a few years after the item was manufactured. We have come a long way from the days when an oven would last for decades without a visit from the repairman. Perhaps consumers can afford to replace major appliances every few years; the environment cannot afford the associated costs of such waste.

The present-day preoccupation with acquiring things flies in the face of millennia of religious teachings against greed and the desire to "have more stuff." For some religions the accumulation of wealth was viewed positively, it was counterbalanced by the need to be aware of the needs of the poor. The pursuit of wealth without concern for the plight of the poor was condemned. For Muslims, those who wish to neither grieve nor live in fear are those who do not covet to own any wealth at all and earn only as much as they need for their livelihood.

Christian literature is most critical of wealth accumulation, and is replete with warnings about the evils of accumulating wealth—some of the best-known examples are:

- "Be on your guard against all kinds of greed; for one's life does not consist in the abundance of possessions."
- "Do not lay up for yourselves treasures on earth; those who desire to be rich fall into temptation, into a snare, into many senseless and harmful desires that plunge people into ruin and destruction. For the love of money is a root of all kinds of evils."
- "It is easier for a camel to go through the eye of a needle than for a rich person to enter the kingdom of God."

Thus, we see an evolution in the perception of the good or evil of wealth accumulation. In the earliest, rather nomadic times, the accumulation of wealth was viewed as not being such a bad thing. But all of the world's major religions, to which billions now profess adherence, admonish us to not be greedy and to share what we have with needier people. The more recent of the great religions admonish us to not aspire to the accumulation of wealth as it could be viewed that such accumulation is at the expense of the poorest among us. Unfortunately for the environment, religious instruction on this aspect of human life has been overcome by commerce, commercial goods, advertising, and greed. Many professed Christians and Muslims have accumulated wealth beyond reason. On the consumption battlefield, the seductive power of advertising has defeated moral guidance.

As human greed overcame religious instruction, the environment was deliberately destroyed as more and more unnecessary stuff was fabricated and thrown out. In fact, a very large portion of the things we buy are deliberately made to be thrown away. They are what we call "disposable." Few people in the rich world wash diapers and hang them on the line to dry. They buy disposable diapers now. We have disposable razors, disposable cups, disposable plates, disposable straws, disposable napkins, disposable fashion, disposable…

Of course, when a manufacturer creates something that wears out too fast, is unrepairable when it breaks down, or is made to be deliberately thrown away after a single use, that manufacturer does not bear the cost of recycling or otherwise disposing of the waste it has created. That manufacturer bears no responsibility for the damage to the oceans resulting from the discarding of billions of bits of garbage every day. That cost is passed on to the life forms trying to live in the oceans, or to the consumer, and it is passed on in a way that is made almost invisible to the consumer.

Consider the purchase and disposal of a coffee capsule, the disposable container of a single serving convenience coffee preparation. Let's assume the purchaser lives in a downtown apartment and works in an office nearby. The cost of each coffee capsule includes the cost of the plastic container, the foil lid and the coffee therein. (The coffee is likely the least expensive of the components.) The consumer buys the product, uses it once, and generally throws it away. Someone picks up the garbage, perhaps sorts it for recycling, transports it to a facility for recycling or disposal, recycles it, ships it off to a third country for disposal, or buries it. All of these activities have costs. Let's say the consumer paid $0.75 for the capsule. The local municipality pays the cost associated with the waste and adds that cost to property owners' taxes. If you are an apartment dweller, your landlord pays the taxes and includes a portion of those taxes in your rent. The person who owns the office building does the same. The consumer has no idea of the environmental damage from untreated coffee capsule waste, or the actual cost of preventing such damage by the thorough recycling or treatment of such waste.

Imagine a world where there is a ban on transporting any and all disposable consumer items off your property. The apartment dweller would very quickly be out of space. The homeowner would rapidly fill up his house and all the land around it. We don't do that. So, we pay someone to dispose of our waste for us, but have little idea of the cumulative cost of doing that. Now imagine a world where there is honest advertising of the costs of a cup of coffee. The coffee shop would have to charge $2.50 if it included the environmental cost. But it only charges $1.50 and leaves the costs for waste disposal and all other environmental costs to others. The environmental cost of the coffee-shop coffee is kept from the consumer in the same way it is for the coffee capsule consumer. Coffee capsule plus environmental cost: $1.50 per cup. Coffee capsule without consideration of the environmental cost: $0.75. Brew your own from ground coffee: $0.15 per cup. These are rough estimates, of course, but the concept is clear. When you have a cup of coffee, you have no idea of the total cost of dealing with the waste that has to be dealt with as a result. In hiding these costs, we are wasting the environment by design.

Recreation

Imagine you are out for a day's paddle in a canoe. You are disturbed by a noise coming from a couple of kilometers away. At 1,000 meters you can hardly hear your own voice. Your recreation is ruined for the time being. The quiet enjoyment of the environment for all life forms nearby is ruined.

What happened to the quiet, healthy days of our youth when the most favoured form of recreation was a walk in the woods? Kids played ball-hockey, rode their bikes, and built forts in the bush. Teens would pack their kit and head into lake country for a week of paddling and portaging. Families looked forward to a once-a-year camping trip, or if they were lucky, a week in a cottage.

Recreation is a broad term. Basically, it means things that are done for amusement, entertainment, or pleasure during one's leisure time. Such activities are intended to be enjoyable. Recreation may be playing golf, tennis, or bridge with one's friends; healthy and fun activities that improve mind and body. It may mean a quiet walk in the woods, paddling a canoe or going for a bicycle ride on a sunny afternoon. Going out to a concert of live music or sitting down with a good book are other forms of environmentally friendly recreation. There are hundreds of different forms of positive, healthy recreation. Unfortunately, there are now thousands of different forms of unhealthy, environmentally abusive, recreation.

Consider the roar of an Indianapolis 500 car race, for example. Or the Grand Prix de Monaco, the fumes from the engines polluting the atmosphere and shortening the lifespan of those sitting alongside. These are the modern version of the old-fashioned horse race. At least with the horse races there was a meaningful sense of competition between magnificent well-bred beasts and their fit riders. Now it's a bone-jarring competition to see which group of engineers and technicians have put together the winning vehicle, while the audience hopes for a few spectacular crashes along the way.

Then there is the Monster Truck Chaos. A weekend of fun for the family where huge vehicles roar around crushing other vehicles and filling the air with fumes, as the local media points out what a wonderful thrill this is for the kids. Just what message are the kids getting? That good family recreation is to sit in a fume-filled stadium watching things being destroyed? Of course, if one wants to enjoy a form of recreation where the objective from the start is to demolish things, then where better to go than to a Demolition Derby Smash Up. Just see how many vehicles can be demolished in a single afternoon. What fun! You get the message. Inspired by such Monster Mash and Demolition Derby experiences, today's youth head out in an oversize pickup truck with a couple of trail bikes in the back for "recreation" in an idyllic forest. They drive through the muck as far as they can go, fire up the trail bikes, and roar through the trees. What fun! Never a deer seen, or a bird heard. Native wildlife is frightened and flees the area until silence returns. In winter, trail bike riders swap their trail bikes for snowmobiles. No crunch of snow under foot. No exercise from poling along on cross-country skis. Pity the poor skier who wanted to get out for some real recreation and has his day destroyed by thoughtless machine drivers.

Recall the cigarette boat with which I began this chapter. The biggest among them can burn 320 kilograms of fuel an hour, emitting more than a thousand kilograms of CO_2 as they do so. The noise disturbs hundreds of residents along the shoreline and all wildlife in the area. This is recreation? And is it recreation to gather by the thousands to watch a competition between such boats? Does anyone give a thought to the fact that in so doing they are watching deliberate environmental abuse? To the contrary, electronic media, especially television, glorifies and encourages this kind of environmental abuse. They broadcast many such events and play the crashes over and over again. This is just part of the damage done by media outlets every time they glorify a pie-throwing contest, monster truck crash, speed boat race, and so on. The local news should not praise the homeowner with the most garish display of bright lights and seasonal inflated lawn ornaments; they should call out these environmental abusers for what they are.

The ultimate "vacation" experience now is a million-dollar trip to space in a rocket ship. Some people laud the ability to throw a two-hundred-pound man into orbit for a few hours or a day, few lament the fact that in so doing millions of pounds of precious fuel are thrown away and the environment is defiled in countless ways. One of the problems with the recreational habits of the super-rich is that many choose to spend their bloated share of global resources on wasteful pursuits—huge yachts burning a couple of hundred liters of fuel each hour, or a trip into space. As the super-rich would say, "It ain't cheap, but the smile factor is great." Who cares about the environment? The rest of us will pay the long-term costs.

Closer to home, recreation can be as simple as going to the movies. While moviemaking may not be too destructive of the environment, one must question the message being transmitted to the movie-goer. All too often, movies show battles between opposing forces where the objective seems to be to destroy the largest amount of material in the shortest period of time. Cars explode, bodies explode, buildings explode, planets explode. What is the message? Wholesale destruction is entertaining.

How can the current generation grow up to love and care for our planet?

Part III

SOLUTIONS FOR A WOUNDED PLANET

The first part of this book described the present state of the environment, particularly the environmental commons of air, water, land, and species shared by all. It showed that the environmental commons has been seriously degraded by human actions over the last few centuries. Part II, "Wasted by Design," examined human activities and how they had been designed to damage the environment while maximizing profits and wealth accumulation for the present generation. Now we will look at some of the thousands of solutions for this wounded planet.

These solutions begin with the actions of individuals. Every individual can play a part in healing environmental damage. More can be achieved by working with family and friends, neighbours, and community. Communities working together can bring change to local and regional governments. If enough municipal and regional governments speak out, Provincial or State governments will have to listen. National governments can then be persuaded to change, and they can bring about change at the international level.

There are powerful forces at work on the other side of environmental issues, and they have persuasive arguments to make for their self-interest. Those forces can be intimidating. If enough of us recognize the damage done, and how it was done, we can collectively bring about the changes needed to heal the planet, no matter how intimidating commercial powers appear to be.

This part of the book sets out the steps needed. It is structured in order of increasing global effect but decreasing personal control over implementation.

Actions at the personal level

STEP 1. DON'T DESPAIR!

Choose hope over despair. There is so much that can be done at the personal level, and then at subsequent levels up the chain of possible solutions. Whether you are aware of it or not, millions around the world, if not billions, are on your side in this battle. With their help, we have already travelled a fair distance down the path towards healing the wounds of the past century. In some cases, the forces aligned against environmental healing have pushed us back, but in other areas significant progress has been made despite resistance. But if we throw up our hands in despair and cry, "It's hopeless!" then that will be a self-fulfilling prophesy.

STEP 2. THINK ABOUT THE ENVIRONMENT IN EVERYTHING YOU DO

Rather like our prehistoric ancestors did. They had to be attuned to their environment all the time to survive. We must re-learn how our everyday actions affect the environment and, understanding that, modify our actions to minimize negative effects. There are a million ways to do this, too many to list here. Everyone's daily routine is unique, and the opportunities to act in ways less damaging to the environment are unique to each person. Unfortunately, we have become so disassociated from the natural environment that we seldom think about how each step we take in our daily routines has an effect.

Here are a few tangible examples.

Think of someone preparing a cup of tea. She fills the kettle from the tap and plugs it in. While the kettle is coming to a boil, she runs the tap water until it is hot and fills the teapot, so it won't cool the tea. She empties the hot water, adds tea, and pours the boiling water into the teapot. When it has steeped, she pours her cup of tea, and eventually discards the leftover portion.

To do the same thing, an ancient ancestor might have walked a few kilometers to get a pot of fresh water. She would have carried the full pot on her head back to her campsite. She would have collected some wood—a lot of wood—to build a fire to boil a full pot of water. She would have used some of that to warm another pot, added her tea leaves and a considerable amount of water, had her mug of tea, and thrown the rest away. Of course, being environmentally aware, this ancestor wouldn't have done that at all. Knowing how much work was involved in getting water and fuel, she wouldn't have wasted anything. If she had a hot drink at all, it would have been made from the minimum amount of water necessary, probably heated over a few hot coals remaining from cooking supper.

Now, thinking about the environment doesn't mean going back to the old days. It means thinking about everyday things we do in terms of how our actions affect the environment. In this case, if we want a cup of tea, why heat more than a cup of water? Learn how much water needs to go in the teakettle to make one cup. If you had to go out and collect a few twigs to heat a cup of water, you wouldn't be inclined to gather an armful, just so that you could throw away excess hot water. Similarly, you wouldn't be inclined to waste your energy and precious time warming a pot to contain the excess tea. Thinking about the water in your cup of tea; it had to be collected somewhere, by someone. It had to be treated to make it drinkable and transported, generally by pipe, to your residence for your use. All these steps take materials that wear out and use energy, while depleting a precious resource—water. If one is going to make just one cup of tea, do it in the cup.

If you live in a house, when you do the laundry do you automatically throw the clothes in the dryer? Or do you use the solar and wind-powered clothes dryer and fabric freshener (that would be a clothesline), if available, and when weather permits. Most people don't even think about the cumulative environmental damage associated with the unthinking use of machinery, perhaps because they feel that living *in* the environment and *working with it* is just too much work. If they were to think about this, not just in terms of the negative environmental effects, but rather in terms of the positive benefits of getting out in the environment and having it help them, they might use a clothesline more often. Don't think of it as a work penalty—think of it as a chance to be in the environment, feeling the sun on your face, hearing the sounds of the world instead of being in front of a video screen while a clothes drying machine rumbles away in your basement.

The problem is, we have become so disconnected from the environment, the source of our water and energy, we don't even think about how our actions abuse the environment.

Here is another example. You are planning the evening meal and you realise you've run out of potatoes. You hop in the car, drive to the grocery store, and pick up a bag of potatoes. Thinking of potatoes, you notice a display of yummy-looking potato chips, so you throw a couple of bags in the cart; and maybe a couple of cans of pop to go with it. Back to the car, back home. Oops—cooked too many potatoes; kids shouldn't have had those chips and pop. They didn't finish their supper. Oh well, potatoes are cheap, just scoop the leftovers off the plate into the garbage. Nary a thought of environmental abuse along the way.

If you had to plant and harvest your potatoes, you might have thought ahead—well ahead. You certainly wouldn't have wasted any; it would have been too much work to go out and gather more for tomorrow. That display that seduced you to buy the potato chips? Well, someone had to mine the bauxite ore from which the aluminum bag liner was made. That is a very polluting and environmentally damaging activity. The refining of aluminum from bauxite is very energy intensive. Production of aluminum pop cans also produces a lot of pollution and consumes a great deal of energy. That pop can your child used was just one of the 7,000 cans used around the world each second. String them together, and in one day you would have a string that goes all the way around the world almost twice. If you had to mine the ore, transport the ore, refine it, fabricate it, and deal with all the mess left over, you might think twice about putting pop in cans. The same line of reasoning applies to the brightly coloured plastic wrap overlying the aluminum foil potato chip bag. It came from oil, which had to be dug up from the ground, refined and… Well, you *know* the rest of the story. If you had to do all the work of creating that package and saw all the environmental damage you left behind as you did so, you'd think twice about putting a few ounces of potatoes in it.

That's the real message here: *we need to think* about things that the consumption-driven commercial world doesn't want us to think about.

People are encouraged to set back their thermostats in winter and reduce air-conditioning costs in summer, and these are good things. But one needs to think about the less obvious things as well. For instance, how often do you open the fridge door to get out a single item, return it, and then a few seconds later open the door again to get another item. Even worse, leave the door open for a minute or two while you remove, use, and return several items in succession? If you thought about the environment at every step, you would change your behaviour. You might think about all the steps required to create the energy that runs your fridge; all the steps that are required to get that energy into your house; how your behaviour affects your own home environment.

That's my message here; think about the environmental effects of your everyday actions at every stage. Think about the environment until it becomes a comfortable, second-nature kind of thing, as it was for our ancestors and as it should be for us.

STEP THREE: ACT

Here are a few things that you, as an individual, can do to help heal this wounded planet. These things are under your control. More than any other actions, you can make them happen. Each individual action may be small, but enough people, doing many individual things, will have a large impact. Nobody's perfect, so don't despair if you can't do everything possible to save the planet. Lots of people doing lots of things will be better than where we are headed now. Overall, these actions can be summarized as:

> Don't eat so much.
>
> Don't travel so much.
>
> Don't consume so many things.
>
> Don't breed so much.
>
> Do enjoy the beauty and tranquility left to you in this world and try to leave more for the next generation.

Food

First and foremost, review your eating habits and revise as appropriate. Overeating and eating the wrong types of food place a huge burden on the environment. Plan ahead for food needs and avoid unnecessary trips and buying more than necessary. Food waste is a terrible environmental problem. It is better to finish a meal with an empty table than to leave that table with food to be thrown out. We eat more than necessary for good health anyway, and the environmental cost of producing surplus food and then throwing it away is a terrible, deliberate assault on the environment. This is entirely under your control.

Think about what you buy for food. Do you or your family really need chips? Do they really need soda pop? The producers and marketers of food have learned that we crave foods with sugar, salt, and fat. This craving originated in ancient times because such ingredients were helpful for immediate energy (sugars), necessary trace minerals (salts), and long-term stored energy. There is nothing wrong with these ingredients themselves;

it is the quantities we ingest that harm our health and waste the environment. Food producers and marketers don't pay our health bills or the cost of environmental damage, but they certainly profit from our over-consumption of things they entice us to eat. Try to reduce your intake of animal protein per person to the amount needed for good health. Above all, avoid over-processed and over-packaged foods.

Avoid foods produced with excessive hormones, antibiotics, and pesticides to the extent possible. If you can grow some of your own fruits and vegetables, you will be in charge of pesticide use yourself. You can decide to manually remove tomato worms and potato beetles, instead of dusting them with insecticides, for example.

Some food waste is unavoidable; peelings, roots, vegetable tops, and so on. So, compost them. Hopefully you have a garden in which you can use compost. If you live in a house with a yard and don't have a garden, put one in. Plant a fruit tree or two instead of an ornamental bush. It will give you pleasure to have your own produce, prevent environmental damage from food transport, and help you to connect to the real environment. Most properties have some place that can be turned into a garden. If you can't use the compost yourself, give it to a friend who can, or look for community organizations that collect compost.

Avoid buying things wrapped in plastic, especially things double-wrapped or triple-wrapped. If you can't avoid the plastic wrapped and overwrapped items found almost everywhere, let the vendor know that you are unhappy about paying for unnecessary wrapping and even more unhappy with having to pay to recycle or dispose of the unnecessary wrapping. If circumstances permit, without the rude interruption of those around you, remove excessive wrapping and leave the waste with the vendor for disposal. If enough retailers get the message, they might transmit that message to wholesalers and thus to suppliers. Ideally, higher levels of government would take action to transfer the cost of unnecessary waste disposal back to the source. Until that happens, individual purchasers may have to take matters into their own hands.

Avoid buying bottled water. Much of the world has safe, clean water that has been produced and monitored at great cost for your safety. Even in such places, too many people carelessly use water that has been carefully collected, purified, stored and distributed. They use a lot of it for watering lawns and washing cars, while ignoring the fact that it was purified at considerable cost for cooking and drinking. The environmental cost of producing good drinking water is what I would call a necessary cost. Using bottled water instead is an unnecessary assault on the environment. Humans use more than a million bottles of water each minute. That's about 1½ billion plastic bottles a day—more than half a trillion plastic bottles a year. Most of those bottles wind up in the ocean or in landfills where they may take hundreds of years to decompose. Probably less than 20% are recycled. It takes about 2,000 times as much energy to produce bottled water than to produce tap water and consumes non-renewable resources along the way. Most bottled water contains microplastic particles. Why buy it if you have secure potable water in the tap?

Bottom line in the nourishment department: eat less (you will be healthier and save the environment at the same time); avoid highly-processed foods; reduce your meat protein intake—a quantity about the size and thickness of the palm of your hand (not including fingers and thumb) is sufficient for good health; don't drink bottled water.

Shelter

With respect to your clothing, the immediate shelter of your body, avoid fast-fashion items, recycle clothing to the maximum extent possible and reuse second-hand apparel where possible and appropriate.

Do an environmental audit of your home to see if there are things you can do to improve its environmental friendliness. You will find plenty of information resources specific in your locality to help with this. Do you have multi-pane windows? Is your insulation adequate? Is your furnace efficient? Is it possible to install a high-efficiency wood-burning stove? The answers to these questions and the implementation of solutions could be quite beneficial to the environment. Thinking a little more broadly, would it be possible to establish a vegetable patch or plant a few fruit trees? Good for body, spirit, and the environment, all at once.

You just need to think about your shelter from an environmental perspective and do the right things.

Security

Most environmentally damaging actions with respect to security happen at higher levels of social organization. Your most constructive role here may be working in concert with family, friends, communities, and local governments to shift higher levels of government away from military solutions to perceived security issues and towards negotiated solutions to security issues. The resources devoted to the planned deliberate destruction of resources are a designed waste that destroys our environment in so many ways that they defy comprehensive description.

On the personal level, you can choose not to rely on weapons for personal security and work with others in your community to create gun-free areas. By setting an example and discussing the environmental reasons for doing so with your friends and neighbours, you may be able to expand such action to higher levels of government, in recognition of the group safety benefits of such action.

Health

On the personal health front, in addition to reducing your food intake to both save the environment and become healthier, stop using unnecessary antibiotics. You don't have to kill 99.9% of bacteria on your floors and in your toilets. They'll be back, and more antibiotic-resistant than before.

Focus your personal health care on preventive measures rather than responsive measures to the extent possible. When responsive measures are necessary, and such measures involve medications, be sure to return left-over medications to the pharmacy for proper disposal. Medications thrown down the drain or into the garbage find their way into the environment where they cause widespread damage.

In your recreational activities, choose things that use your own muscles instead of machines: walking, not riding an ATV; cross-country skiing, not snowmobiling; kayaking or canoeing, not powerboating and, so on. The "personal energy" forms of recreational activity save the environment, keep you healthier and lessen the burden on health care systems.

Energy

Think about your energy use in the home, not just in adjusting the thermostat but in everything you do that requires energy input from an external source. Some things, like how you make a cup of tea or how you use your refrigerator, are obvious, but there are thousands of things you do each day that take their toll on the environment because they use energy unnecessarily. You need to do your own analysis. Did you close the door as promptly as possible on a cold winter's day or did you stand there chatting with guests as they were leaving, without a thought about the amount of heat pouring out the door and the environmental cost of replacing it?

Transportation

This is likely the most significant area under your direct control where you can have a significant effect in reducing abuse of the environment. Reduce your travel and when you do have to travel, do so in the most efficient way, from the smallest trips around town to the longest, around the world. If you can combine two, three, or more trips into one—do it. If you can use a bicycle instead of a car for a given errand, do it every possible time. Remember, cycling is about twice as efficient as walking, and many times more efficient as any other form of transportation. If you have good public transit, use it as often as possible. If you want a holiday, search for pleasure closer to home—think about the environmental damage associated with long-distance vacationing. You have absolute power to choose not to travel to the other side of the world for a vacation.

Of all the possible ways to help the environment, suggesting that people travel less frequently is the most likely to elicit an immediate negative reaction. Often the response to the suggestion that less air travel would be good for the environment is: "Well the planes are going to fly whether I'm on them or not!" If everyone adopts this attitude, then it will truly be a self-fulfilling prophesy. But if just half the people who fly were to do so half as often, the number of flights would drop by one-quarter. If half the people who fly unnecessarily were to just stop flying then there would be 50% less pollution associated with building airplanes, pumping their fuel out of the ground, and burning carbon to keep them aloft. Add the reduced need for the construction and maintenance of airport facilities and the environmental benefits grow larger.

People need to look for alternatives. If you are invited to make a presentation on the other side of the world, transmit a video instead. Even a professional video is much less expensive than an air flight, and so much easier on the environment. There are many such alternatives. In business, they might include high quality videoconferencing, which is almost like being there. For seeing foreign attractions there are now virtual reality tours. Of course, people don't get bragging rights for such alternatives. But if bragging rights outrank healing environmental wounds caused by excessive travel, the planet will not recover.

When you move to a new job, try to find a place to live as close as possible to your place of work; preferably within walking or cycling distance. The exercise will be good for you, it puts you in touch with the real environment, and saves time and money by reducing commuting needs.

Durables

This is difficult at the personal level. People need to assess everything they buy from the point of view of durability, rather than style or the latest fad. This assessment should extend to everything from toothpicks to automobiles. You are up against pretty stiff competition here; commercial interests are going to try to persuade you to buy more than you need, and replace it more often. They will seduce you to get "two for one" whether you need two or not. You are not really getting two for one. You are getting two items at what is probably a reasonable price for each item, times two, packaged to look like you are getting one for free. There is no free lunch.

Commercial seduction to buy things has been perfected to a very high level. Look beyond the seduction. Are you really buying a car to smash through snowdrifts or ford streams and plough through muck? Are you buying that car because if you do, a beautiful young woman will find you irresistible? No, you are buying that car to get yourself from A to B in the most efficient way possible (assuming you can't go by bicycle or public transit). That will be the way that is least damaging to the environment if you must travel by car.

When you need to replace a large appliance, do your homework. Look for the one that has proven to be most reliable with lowest repair record, not the one with the most gadgets incorporated: gadgets that will inevitably break down, rendering the entire device useless.

Buy the best quality tools you can afford. Buying cheap tools almost always turns out to be the most expensive option, as you find they need to be replaced often, instead of lasting a lifetime. My favourite tools are the ones my Dad passed down to me. This applies to tools needed in the kitchen as well as the garden or workshop. Every tool thrown away prematurely is a double insult to the environment; the primary resources consumed in its production and the premature burden of waste disposal.

Recreation

Choose human power over fossil fuel power when you want to get out and enjoy yourself. Cycling, walking, jogging, and hiking have very low environmental impact and great health and spirit-uplifting qualities.

Explore your local community for everything from dining out to seeing some beautiful sites. You don't have to travel to the other side of the world.

Think about the environmental implications of maintaining two residences that allow one to flip back and forth from season to season. Canadians call people who enjoy this lifestyle "Snowbirds," and there are organizations dedicated to maximizing benefits for those who choose this lifestyle. The environmental implications are significant: pollution associated with travel between the properties, providing utilities for unused houses, removing property from the pool of housing needed by others, and so on.

Other things you can do

Get involved with organizations working for the repair and preservation of the environment. Choose political parties at the local, regional, and national level that are most directed towards environmental action. Vote for those parties and support them wherever possible. If you are an investor, direct your investments towards environment-sustaining enterprises like renewable energy companies, and away from damaging ones, like fossil-fuel companies.

If you are a person who donates to charities, think about the charities you choose with an eye to the need for support of environmental organizations. Some beneficiaries will be obvious, others less so. You may support health care and housing projects in needy areas around the world, but do you remember the need to support the education and empowerment of girls and young women so that they can help to bring global population under control. Remember, the population explosion resulting from irresponsible human reproduction is the single biggest cause of environmental damage. Some governments deny support to foreign aid work if that work includes family planning actions—you, and like-minded individuals, will have to fill that funding gap.

These are but a few of the millions of personal actions you could take to heal the planet. It is not possible to list all such actions here, because everyone's situation is different. Don't beat yourself up when you discover that some of the things you do, and like to do, are not the absolutely best environmental alternatives; it is a question of balance—as we shift the balance in favour of the environment, things get better. Nobody's perfect, but we can all take one step at a time in the right direction.

The important concept at this level of action is to think about what you do from an environmental perspective, and act accordingly. Every action, from turning on a light switch to having another helping of meat at dinner has an environmental impact. You will have to educate yourself to learn what the environmental effects of your actions are. Think about children being driven to school, instead of walking or cycling; workers driving to work instead of using public transit; food coming from grocery stores instead of gardens; trips to resorts on the far side of the world instead of camping in a local park; being so distracted by mobile devices that one becomes oblivious to the environment around them…

We have become so disconnected from the environment upon which we depend for survival that it is difficult to figure out the totality of the effects of our actions on the environment. It may difficult, but it is worth trying: the planet will not heal unless we do.

Remember, if only half the world's people do what they should, that's almost four billion people helping to restore the environment. That makes for a big step in the right direction.

Actions at the Family level

Let's assume you have decided to take control of your own actions to reduce environmental abuse. You have greatest control over those actions. Hopefully, you have decided to not despair, to think about what you do in terms of your effect on the environment, and to act in ways that minimize the environmentally damaging effects of things you must do to live. The next step, over which you have some significant measure of control is with your family and friends.

You may be pleasantly surprised to discover that your children, parents, grandchildren, and friends already know more about how to heal this wounded planet than you do. Sometimes people fail to discuss the environment in social settings and may be unaware of how their friends and family feel about it. You may find that your family and friends are very well informed on environmental issues and may have some great ideas about what can be done as a family or a group of friends to help. If they are not so informed, you may be able to gently, but persuasively, help them turn their energies in the right direction. Actions will speak louder than words. If they see you deliberately not buying a bunch of stuff to make a temporary Halloween or Christmas display, they may conclude that you are sincere in your efforts to heal the planet. Same thing with the kinds of gifts exchanged. Do you rush out to buy boxes filled with the latest faddish plastic toys or do you try to find a more meaningful way to interact with your children and grandchildren? Is an energy-consuming trip to a noisy, flashy, midway show a meaningful interaction with family? How about a pleasant walk through the community park instead?

If you must make a trip to the nearest shopping centre, check with your nearest friends to see if there is anything you can pick up for them. It not only lets them know that you are willing to save them a trip but signals that if they were to do the same for you, the environment would benefit.

The most important thing here is to share knowledge and experience with your family and friends about ways to heal the planet. Let them know when you think they are doing great things for the environment. Ask for their ideas when you have an environmental problem that needs attention. All members of the family need to be aware that things like on-line video games and shared videos have their own energy-environment impact. (Remember all those internet server farms burning up as much energy as a medium-sized country.) There may be things you can do collectively, like helping to restore local streams or cleaning out invasive weeds in your community. As was the case for personal actions, there are probably a million ways you and your family or a group of friends could do good things for the environment because every family and group are different.

The challenge is to avoid being overly zealous. Being too single-minded and outspoken risks turning friends and family off. Better to be well-informed on environmental issues and use that information, bit by bit, as circumstances permit. For example, you may have friends who still do not believe that our climate is changing because of collective human activity. Their beliefs may stem from a political bias that chooses to deregulate commerce in favour of near-term economic benefit. If that is the case, you may have to walk away from the conversation and hope that the subject can be revisited again later. This may be most challenging with respect to population control issues. If you see that someone is going to refuse to accept the idea that interventions that might save the environment by preventing excessive births, on a matter of principal, you may have to step away from the discussion, even though you know that overpopulation is the single largest threat to the environment.

Here are a few things you might talk about with your family.

Food

In the same way that you, personally, have decided to not overeat, and avoid eating the wrong types of food, set an example for family and friends and encourage them to do the same. When you have a family gathering for a meal, plan ahead to minimize leftovers and then plan ways to use them up completely. It is horrible to see hosts scraping great quantities of food off plates after a meal, simply because they set out too much in the first place.

See if you can establish a bit of a family food growing and sharing system. Someone might be good at growing fruit, someone else may do well with root vegetables. Anything you can do to provide home-grown food for your extended family and friends will benefit the environment by reducing transportation, packaging, and marketing damage resulting from bringing food in from afar. If a natural disaster threatens, are your family and friends prepared, collectively, to ride out the storm? Or will you be one of the many who rush off to the local grocery storm to deplete the shelves of all their bottled water? Why not maintain a reserve of good old tap water? You may even have a bit to spare for family and friends. The mad rush to buy up all the bottled water when disaster looms is a waste of money and damaging to the environment. Your tap water is probably safer, cheaper, and more likely to be in abundant supply when disaster looms.

Think about your family's use of bottled water. Is it necessary? We have been persuaded by advertisers to believe that we must buy bottled water or, at least, purify our tap water further with some sort of in-house water purification system, regardless of the purity of the water in the tap. This is less a need for better water quality than the desire of manufacturers to profit from selling things.

Shelter

As was the case earlier in this book, "shelter" includes protection of the body from the elements, so we begin again with the clothing issue. Avoid gushing over a family member's latest fad fashion. Encourage everyone to use the clothes they have until clothing reaches the end of its useful life. Accept the exchange of clothing, especially the hand-me-down cycle that works so effectively as children are growing up. Discourage children from demanding the latest style of running shoes or jackets. Talk to them. In many cases, they will not have thought that unnecessarily updating their clothing has an environmental cost. If so, you can take them through the life cycle of producing the necessary materials, sewing, assembling, packaging, transporting, and marketing clothing. These are all things that take a toll on the environment.

Share with family and friends all the ways to make homes as environmentally friendly as possible. Minimizing energy waste, winter and summer, for example. Discourage extravagant displays of energy-consuming seasonal decorations. Don't applaud friends for buying a second house in some faraway place. You may not be comfortable explaining why that is a bad idea, but at least you don't have to cheer them on.

Health

Do you talk with your family and friends about how having more children than needed to sustain humanity places serious stress on the environment? This may prove to be the most difficult topic to discuss as a solution to global environmental problems, especially as the discussion moves from family planning to contraception or abortion in cases of unintended pregnancies. If you have the courage, start the conversations with your children and siblings gently, and see how far you can take it without damaging family relationships. Try discussing it with friends. If they are close-minded on the subject, you may have to just let it ride, and try instead to make gains with them on environmental protection in other areas. At the very least, don't cheer those who rave about the very large families they have created.

While this may be the most sensitive of the environment and health issues, there are many other positive steps that family and friends can take that are not so loaded with political or religious overtones. Focus on them instead. For instance, if a child or grandchild is ill, avoid rushing off to ask your doctor for antibiotics as a first resort, and work with your doctor to avoid the over-use of antibiotics. Help your friends and family to understand how antibiotic cleaners often harm the environment, having no lasting benefit while they contribute to the development of antibiotic-resistant bacteria.

Transportation

Be reserved in your praise for friends who show off yet another new car. Help them to understand the environmental implications of such consumerism. When friends and family rave about their latest "once-in-a-lifetime" trip to the opposite side of the planet where they trampled on yet another wonder of the world, do you nod in appreciation and yearn for the opportunity to do the same? Or do you gently let them know that you have concluded that the environmental cost of that kind of vacation has become too great and you have found local alternatives? Perhaps they might think along similar lines.

Beware: this is the second most sensitive area of solutions for a wounded planet. Many take criticism of extensive travel as an attack on what they perceive as an inalienable right to go anywhere they want, anytime they want, so be careful. Start with a gentle, general question about the toll that global travel takes on the environment. The chart on page 156 might be a good starting point. It shows that transporting a passenger in an airliner produces two hundred grams of CO_2 per kilometer. A trip from Toronto to Paris and back is more than 12,000 kilometres. Your friend and a partner together are responsible for 400 grams of CO_2 per kilometer. That's 4.8 tonnes of CO_2 added to the atmosphere just so they could get to see Paris again. Two people flying to Fort Lauderdale from Toronto and getting on a cruise ship through the Panama Canal to Acapulco, and then flying back to Toronto, would add almost seven million grams of CO_2 to the atmosphere (2.8 million grams of CO_2 by air and another 4.8 million grams by water). That's a pretty hefty air pollution burden for a couple of weeks in the Caribbean, isn't it? Friendly conversations in this area may not be easy or comfortable.

When doing something with friends or family, like going to a concert, shopping, or a walk in the park together, always seek out ways to carpool if public transit is unavailable. Collectively, you can reduce your carbon footprint significantly.

Durables

When you find something particularly durable, share the news. I am reminded of this every time I pick up the under-the-sink compost bucket that I have had for decades. It is made of stainless steel and looks as good as the day I bought it. I expect my kids to inherit it and use it for a lifetime as well. The idea here is that, collectively, you will find ways to multiply the effect of actions that you take individually, and the environment will benefit as a result.

As mentioned in the previous chapter on personal actions, you want to lead your family and friends to think about the environment in all they do, as our ancestors did for their survival. We are no less dependent on the environment than they were; it is just that big business has done everything possible to hide that dependency from us, our friends and family, to increase profits. To address this problem, try to think of putting the words "*so I am going to ruin the environment and*" into a lot of your consumption thought processes. This is especially important in the wealthy parts of the world where efforts to distract us are so great.

A few examples might help:

> "I can afford it, so I am going to *ruin the environment and* take another holiday on the other side of the world."

> "I can afford it, so I am going to *ruin the environment and* have the sixteen-ounce steak on the menu."

> "I can afford it, so I am going to *ruin the environment and* buy a bigger house."

> "I can afford it, so I am going to *ruin the environment and* get a second (or third) car."

> "I can afford it, so I am going to *ruin the environment and* buy another dress, suit, pair of shoes, whatever, because it is so stylish now."

> "I can afford it, so I am going to *ruin the environment and* buy another, bigger TV so we can have one in every room."

You get the idea. People in the wealthy world almost always leave out those four words because they have become so disconnected from the environment that they just don't think of the environmental implications of their use of money.

Elaborating on these ideas in your mind can be helpful. Take the first example and expand it:

> "I can afford it, so I am going to *take a taxi (burning fuel) to the airport (taking up space and using up many non-renewable resources in its construction and operation), get on a plane (burning lots of fuel), take a taxi to a hotel at the other end (taking up space and using up many non-renewable resources there), drive around the destination (burning fuel), then back to the airport, another flight, taxi home, all the while leaving my own home heated, electronics on etc. while I'm away, so that I can* take another holiday on the other side of the world."

Or:

> "I can afford it, so I am going to *have trees cut down, cattle grazed, air polluted by transporting cattle to slaughterhouses, emissions flowing from a meat-processing factory, steaks packaged and shipped to a restaurant, make myself unhealthy and place a burden on the health care system, along with all the associated environmental burdens as I* order that sixteen-ounce steak on the menu."

There so many other things we can do if we think about those four words more often: "I can afford it, so I am going to *ruin the environment and…*"

Nobody is perfect. We, our families and friends, will not always do exactly the right thing. But that does not excuse us from always doing the wrong things, from an environmental point of view. The more often we do the right things, the more likely it is that we are contributing to solutions for this wounded planet.

Community Actions

Communities are groups of people. It may be a group because of physical proximity, such as a group of houses in a specific area, (neighbourhood community), people who work together (an office community, service community or factory floor community), a group of people who are united in religious belief (church community), or a group of people who share a certain interest (the social workers community, or the gambling community). In the present era, "communities" also encompass virtual communities in which ideas are shared on the internet. For the purposes of this book, a "community" is group of people with whom you associate, beyond family, but short of political groups. Political actions to heal this wounded planet will be addressed in the subsequent five chapters.

The most immediate community is the neighbourhood community. In the suburban context, it is often a defined as a subdivision. In more densely populated areas, it may be an apartment community, a condominium community, or an inner-city core residential area that defines itself as a community. These communities have an important role to play in the protection and preservation of the environment.

Residential communities might do the same sort of things that families and friends do, but on a larger scale. The effects on environmental protection will be correspondingly larger. Do they have a "reuse plan" where they can share things that some have, and others might need? Seldom-used tools might be shared neighbour-to-neighbour rather than each owing his own. In the suburbs, produce from vegetable patches and fruit trees might be shared. If the suburban community lacks space for gardens, perhaps an area nearby can be found for allotment gardens. Some neighbours may have facilities for kitchen and garden compost that others may not have. Similarly, some may have a need for kitchen and garden compost that others may not need. The idea here is to share information and resources in ways that help the environment and use resources wisely.

Environmentally damaging activities might be discouraged by implication or by positive discussion. Instead of raving about how magnificent or extravagant a Halloween display is, one might wonder aloud how much energy it takes to keep it lit through the night, how many resources were consumed in its production how much difficult waste will result from its final disposal. As was the case for friends and families, people seldom think of the environmental implications of such displays when they buy them. The focus here is that people need to think about the environmental consequences of all actions. In a community, more minds can be at work to find more solutions. First and foremost—people must think about the environment as they participate in community activities.

Imagine that you run into your neighbour, John, while walking along the neighbourhood streets in late December.

John says:

> "Did you see my magnificent Christmas display this year. I love to celebrate Christmas. I should win the local newspaper award for best display this year!"

To which you reply, without thinking:

> "Wow, that sure is impressive. You get my vote for best display on the street!"

Thus, reinforcing John's action.

If you think about it, what you *might* be hearing John say, is:

> "Yes indeed, I have wasted a lot of resources and energy to put this together. Isn't it marvellous? Shows you how much money and time I have to spare. Not only that, when it's working, I'm wasting a huge amount of energy every minute. Aren't I wonderful!"

To which you could reply:

> "Truly wonderful, Dick. When we have visitors this holiday season, we're going to bring them over to show how you can really waste time and money and destroy the environment along the way. They'll really be impressed. I read that you've won the local newspapers' prize for the most waste this year. Isn't that awful!"

Okay, these may not be the exact words, but beneath the visible in this scenario lies the reality that when holidays arise some people can't help but compete to set up the most extravagant displays of deliberate waste on their front yard. The same sort of thing happens at Halloween and Thanksgiving.

So, how to deal with this?

There are several possible steps. The first, and most difficult, is the personal level. When you see your neighbour, you tell him you've seen his display and ask him calmly and politely if he thought about the waste in resources that goes along with what he's done. You don't want to make him mad, just to help him think about the other implications of what he is doing.

You could talk with the neighbours on your own street and come to an agreement about the level of holiday decoration that might be appropriate without be so wasteful. You are not trying to be "the Grinch who stole Christmas," just trying to sensitize your neighbours to the fact that there are environmental implications to everything we do.

You and like-minded neighbours might write to the local newspaper to suggest that, instead of encouraging this kind of practice by drawing attention to it, they should write about the negative environmental implications of what is going on. Once again, it's all about sensitizing the population to think more about the environment in everything they do. As was the case when promoting environmental sensitivity with family and friends, you want to avoid coming across as overly zealous, as that is likely to turn your neighbours off, rather than encourage them to help. It is a fine line between positively promoting responsible environmental action and stimulating hostility.

In apartment communities the questions to be addressed may be different, but the concept is the same. Does the apartment building have an easy-to-use and effective recycling system, for example? Does the apartment complex have a bit of land on which residents may have allotment gardens? Is there a bulletin board on which tenants can let each other know about things they can share or pass on to others for reuse? These considerations apply to other residential communities as well. Condominium communities or inner-city communities will have different sets of issues to address, but all can address them in a positive fashion in favour of the environment. As was the case for individual actions and family actions, there are millions of ways in which residential communities can help the environment. They cannot all be covered here, but each community can define its own positive action if the people in that community think about it.

A community often overlooked in this area is the "work community." In most work situations there is a community of workers with a common purpose, whether it be to fabricate a part for an automobile or develop a policy in a government bureaucracy. Opportunities to think about, make recommendations, or take actions to help preserve and protect the environment abound in the work environment. Workers should speak out about ways to make operations more environmentally friendly and should be rewarded for their suggestions accordingly.

On the factory floor, is everything possible being done to minimize waste in the fabrication of things? What about steps to reduce energy consumption? Are there ways in which the production of something can be combined with other steps in the process to allow for recycling of components or overall energy minimization? In the millions of production lines around the world there are undoubtedly tens of millions of ways to help the environment, if the work community would only keep the environment in mind, instead of a single focus on making as much money as possible. Making operations more environmentally friendly in the short term will most likely help the bottom line in the long term anyway.

At the next level up is another community, a community of factory managers who, collectively, can achieve even more for the environment. Whether they are pushed by their employees below, drawn by higher management, or just decide to propose action because it is the right thing to do, they can achieve much in the field of environmental protection. Here again, the first step is to think about the environment in the context of work, and then do something about it. Industrial "communities" need to start switching to low carbon practices sooner rather than later. Every industrial investment decision should answer the question "Is this new facility we are proposing likely to be shut down to satisfy environmental regulations because it is not the best alternative from an environmental point of view?" If the answer is "Yes," then the options should be reassessed to prevent wasting resources on a facility that will be environmentally outdated before it is operationally worn out.

The bureaucrat working on a policy paper should ask how the policy will affect the environment or other policies which could, in combination, be good or bad for the environment. Has the culture of environmental preservation percolated throughout the organization so that no policy is developed if there is an alternative that will achieve the primary objective and offer more environmental protection along the way?

Between these two examples there is a vast array of work environments, from construction, food production, retailing, financial management, public works, and so on. In each of these areas, there will be countless ways to do things effectively and save the environment at the same time, if only work communities would think about it.

Start with the smallest of actions, making the morning coffee in the office. Offices used to have coffee percolators. Then there was instant coffee. Drip coffee with paper liners. One-cup paper filters. Now there are coffee pods and coffee capsules. Does anybody think about which of these alternatives is least damaging to the environment? Coffee capsules are likely the most damaging. They are plastic containers with an aluminium foil seal and a portion of ground coffee, enough for just one cup of coffee, contained inside. There were about 60 billion coffee capsules made in 2018 and 95% of them wound up in landfill sites where they may take up to 500 years to degrade. Most people have no idea of the environmental damage from these items and just throw them in the waste bin, because most people, in offices or at home, just don't think about the environment when they make a cup of coffee.

So, here's a small step, entirely under the control of an office community: take a collective decision that staff will not bring store-bought coffee in the cup into the office, and coffee prepared on-site will be made in the most environment friendly fashion. Everyone saves the environment and coffee money at the same time, and just doing it reminds employees that they can do things to help solve the world's environment. Moving up the scale slightly, what about wearing heavier clothing in the winter and much lighter clothing in the summer and adjusting the workplace thermostat accordingly. There are tremendous environmental (and economic) savings here as workspace heating and cooling have significant environmental effects.

Does your workplace allow the use of internet streaming for personal or business purposes? Internet servers are huge energy consumers. A policy change here can be of great benefit to the environment.

Moving to more significant environmentally friendly steps in the office community, large national and international organizations need to think about the costs and advantages of reducing long-distance travel. Too many organizations decide, without thinking, that because it is time for another big meeting everybody concerned should get on a jet plane and fly to a meeting destination. Choosing a video conference over a physical presence not only saves time, inconvenience, and money; it saves the environment as well Most people just don't think about the environment in the workplace—that's what must change if we are to heal this wounded planet.

Many people belong to religious communities. These communities may or may not have a physical presence: a church community; a temple community; a synagogue community, and so on. Where there is a physical presence like a church hall, for example, is the community taking all reasonable steps to minimise adverse environmental effects? Do people running events at temples synagogues and churches do everything possible to reduce waste, conserve energy, minimize heating and cooling waste, share common goods, and reuse common goods?

Here again, there are more avenues of environmental stewardship open to the many minds of the community than a single author can conceive. So, put those minds to work collectively, for the good of the community and the environment, to find new solutions to environmental problems. Many religious communities support humanitarian initiatives abroad. When they do, they should think carefully about the extent to which those initiatives may damage the recipient's environment rather than improve it. All religions profess some responsibility for protecting the world in which they live and worship; they just don't think about that stewardship responsibility often enough.

Social communities, like golf clubs, bridge clubs, hiking groups, service groups, and so on, also have a role to play. For some of these, like golf clubs, the environmental impacts of their activity will be obvious and severe. Golf was originally a winter game, played in Scotland when the grasses had died down enough to have a reasonable possibility of finding the ball. With the development of mechanical grass cutters, golf became a summer game and spread around the world. Where are the golfers now? They are found on carefully manicured courses all over the world using lots of fertilizer, pesticides, and water which often comes from limited aquifer supplies in the middle of deserts. The dozens of golf courses in the Palm Springs area of California are a prime example. If people want to go for a good walk in the hills, why not? They don't have to damage the environment to do so; they just have to think about the environment before proposing to build yet another golf course.

Hiking clubs seem to be acutely aware of the environment and the need to protect it. This appears not to be the case for many off-road cyclists and ATV (All-Terrain Vehicle) drivers. Hikers frequently see and lament the all too obvious damage caused by mechanized trail users. Here again, those involved in mechanized trail activity often appear to be unconcerned about the environmental effects of their communal activities. This is particularly obvious in the wintertime contrast between a community of cross-country skiers and a corresponding community of snowmobilers. One is safe, quiet, and respectful of the environment. The other is dangerous and noisy—overpowering and damaging the environment. If the objective is to get out in the fresh air and enjoy the natural world, then the mechanized community needs to think of the total environmental impact of their activity and choose not to do it. How is the environment affected by the fabrication and fuelling of a snowmobile? How much is the peaceful quiet of a snowy day disrupted by the roar of a snowmobile? How much is the pristine air quality of the deep woods in winter wounded by the carbon dioxide laden, smelly fumes of a snowmobile?

It may seem strange to think of the communities that include bridge clubs or service clubs as having a role to play in the care of the environment, but they do. If the conversation in such groups revolves around the latest fashion or the last globe-trotting excursion and all are agog at the marvellous ability of some members to spend their wealth this way, then those communities are subtly contributing to the world's environmental problems. If, instead, the conversation revolves around the novel and clever ways people have found to improve the environment, then they are part of the solution, not part of the problem.

The common thread here is the need for thoughtfulness. Because we are so isolated from the real environment in most of what we do, we seldom think about it. Thinking about it should give rise to positive discussions in the many communities of which we are a part. Then, perhaps, those discussions may lead to positive action. Not thinking, and not acting, is not a solution.

Municipal Initiatives

This is the first level of formal political participation in solving the world's environmental problems. Things begin to get complicated here, as political bodies try to balance competing demands from those who put them in positions of power; positions from which they influence the lives of others. On one side are people who feel that politicians should focus on continued growth as the route to prosperity and a better quality of life. On the other side are those who feel that quality of life, of which environmental quality is an important part, is jeopardized by too great an emphasis on growth as the solution. In democracies, those who put politicians in power fall into two groups; voters, some of whom favour higher environmental quality initiatives while others favour greater material wealth, and lobbyists who favour a particular aspect of growth that is often at odds with the preservation of the environment. In non-democratic political systems such as dictatorships and oligarchies it is more difficult, if not impossible, for individual citizen voices to be heard. In municipal politics, and more so in political structures at higher levels, the degree of personal influence or control over what happens is much diminished in comparison with actions at the personal, family or community levels.

The more effective individuals are in demonstrating a willingness to improve the environment at the personal, family, and community levels, the more likely it will be that municipalities will be persuaded to do their part.

Municipalities and all higher forms of government often must balance two competing approaches to the overall well-being of their citizens. On the one hand, it is easy to demonstrate that there could be an increase in wealth and material well-being by encouraging population growth, construction, industry, and other enterprises. On the other hand, it is possible, though difficult, to demonstrate that there will be an improvement in the quality of life through environmental protection and preservation. Unfortunately for the environment, the protection and preservation side of the equation has a fairly long timescale, in contrast to the commercial side, which enjoys a time-scale more aligned with the political cycle. The challenge for those interested in solving the world's environmental problems is to demonstrate that short term solutions are damaging to long term prosperity.

Opportunities for environmental action at the municipal level abound. They vary greatly depending on size and location of the municipality in question. Too often cities behave as though they are linear facilities: food, water, energy supplies, and packaging come in; sewage, waste heat, and garbage go out. Residents live in the suburbs and drive to industrial areas or the city centre to work, and then return home in the evening. Shoppers drive to malls. Children are driven to school.

Transportation

In most municipalities, bad transportation policies cause the most environmental damage. It is as though the only way to get more people to shop and work in a given municipality is to make it easier for people to drive to shops and offices and park as close as possible to those places. That's the easy solution: allow more hectares of parking around malls, have parking spots along each street, make sure each commercial enterprise in the city centre allows enough parking spaces for the anticipated load, and so on. Cities should discourage unnecessary driving, especially the driving of inappropriately large vehicles. Malls and shopping areas could discourage oversized vehicles from parking close to shops by reducing the space available for turning and parking to that which would be appropriate for small to medium cars only. Special areas for oversize vehicles and pickup trucks could be put as far from the shops as possible to send a clear message that driving a truck to the nearest shopping mall to pick up a few groceries is not an option society favours.

The better, but harder solution to transportation problems is to commit to an integrated public transportation system that makes it easy, convenient, and economical for people to get where they need to go. An average car weighs about 1,500 kilograms. An average adult weighs about 70 kilograms. So, each time someone takes a car to the office, they are moving more than 20 times their body mass in metal and plastic, and burning up a lot of fossil fuels along the way. Imagine, if you can, that you could create a bundle of metal weighing 1,500 kilograms that would fit on your head. You want to go to the corner store for a bag of chips? Put that bundle of metal on your head and start walking. Of course, that's impossible, but it is also something that we almost never think about when we hop into the car to go somewhere.

On the personal transportation level, municipal governments should do everything possible to make transportation by bicycle more convenient. Bicycle transport is by far the most efficient way to move a body from one place to another, in terms of energy use. It is even more efficient than walking. (See the chart on page 111.) The side benefits of doing this are tremendous. Fewer cars on the road means cleaner air. More people on bicycles in cleaner air, means more fit people, which reduces the need for public health facilities. For those who cannot use either public transit or bicycles, municipalities should make it more convenient for electric vehicle use, with more dedicated parking spots and power points for recharging batteries on such vehicles.

Municipal governments are the first level of a multi-layered political structure where adherence to planning policies can be important for the environment. This will vary from municipality to municipality, but the concept is universal. For a municipality with a combination of residential, commercial, and agricultural areas under its control, the development of policies in support of environmental preservation, and the subsequent adherence to such policies, is critical.

This is where the biggest problem in solving the world's environmental ills comes into sharpest focus. At all levels of the political structure, from the smallest municipal committee work to the most global international conventions, there will invariably be vested interests opposed to environmental protection because it will harm the short-term economic well-being of those vested interests. Those who want to protect forests for parks or wildlife habitat, or just to keep oxygen in the air, will be opposed by those who want to clear-cut forests to make way for lumber production, mining, farming, suburbs, shopping malls, and so on. All the activities against the environment are premised on the assumption that the only profitable world is a world of constant growth. The equilibrium that sustained life for hundreds of millions of years has been replaced by a disequilibrium over the last thousand years; a disequilibrium in which the easy way to a richer (short-term) material life is to be found in over-harvesting the world's resources, to the benefit of present-day politicians and the developers on whom they often depend, to keep them in power.

Political and environmental time frames are terribly out of synch. Political systems look at time frames as short as a couple of years; environmental ones may be a century or more. This is most evident in discussions about climate change. It will cost a great deal of money now, to get off our dependence on fossil fuels. It will cost much more, a hundred years from now, to cope with the damage we are now inflicting, assuming humanity hasn't done itself in by then. While this environmental conundrum needs more attention at the national and international levels, there is still a role for lower levels of government to play. Revising transportation systems at the local level is one way to help. It may be costly for today's voters, but beneficial for the next generation.

Waste Management

The next important area for municipal action is waste management. Municipalities can provide accessible areas to deliver compostable materials and other food waste that ought not go to landfills. Landfills should be designed and operated to minimize environmental impacts. Municipalities should have in place systems to minimize their own operational wastes and inform residents about how they can minimize wastes. Another crucial step that municipalities need to take is to track the final disposition of wastes that they export as "recyclable." The assumption is often made that such wastes are actually recycled. Unfortunately, too often those wastes wind up being shipped to poor countries, for a fee, but they are then dumped either on land or at sea.

Property Use and Development

Municipalities can provide incentives and penalties in favour of the best use of property. Modest tax relief for suburban properties with orchards or vegetable gardens would send a signal. Apartment or condominium properties should receive the same stimulus. Properties that are left vacant for extended periods of the year should be assigned a vacancy tax proportionate to their environmental abuse. There is only so much land on planet earth and leaving bits of it unoccupied while using precious energy and water resources to maintain it for extended periods is contrary to healing the planet's environmental injuries. Funds generated on the penalty side could be channeled directly into the reward side. Giving prominence to such initiatives in a municipality's publications would be one way to stimulate environmental awareness in the community.

There are areas where municipalities can exert a great deal of control without having to be concerned about interference from higher levels of government. Construction standards is one such area. The issuance of building permits can include requirements that reduce energy waste to a minimum. There should be penalties, including stop work orders, for those who fail to adhere to the municipal standard. There should be incentives for those who volunteer to go beyond the standard; to install rooftop solar power, for example. Municipalities should not restrict their activities in the construction field to residential construction only. Office buildings, malls, factories, and other commercial operations should also be subject to high environmental standards, not just in energy conservation but also in terms of minimizing waste during construction.

Rooftop gardens which help cool the buildings below and provide a little bit of food for occupants should be encouraged. Waste heat from outgoing water should be transferred to incoming water. Materials composted on site should be used for local gardens.

Another area clearly under municipal jurisdiction is the development of official community plans. The positioning of residential areas close to commercial and service areas reduces the need for long trips for work or services. With effective bylaws for minimum water, air, and noise emissions, residential and work areas need not be incompatible environments. Providing for cycle paths and walking routes between home and work or home and services saves the environment from unnecessary transportation pollution and encourages people to get out in the environment and appreciate it more. The result is not only environmental protection, but better health and fitness.

Noise and Light Pollution

Something that often escapes attention in municipal environmental care programs is the need to keep noise and light pollution to a minimum. Unnecessary noises, like unmuffled motorcycles, should simply be banned. Necessary noises from industrial and construction activities should be minimized to the extent possible. Municipalities should have noise bylaws that are regularly reviewed to accommodate the latest best practices. Light pollution is often just a waste of energy and a blight on the nighttime environment. Billboards blazing in the middle of the night alongside roadways are wasteful and would not be cost-effective if those who use them had to pay for the environmental damage associated with their use.

Steps are already being taken in this direction. The LEED program (Leadership in Energy and Environmental Design) provides verification that a building or community was designed and built using strategies aimed at improving performance in energy savings, water efficiency, CO_2 emissions reduction, and stewardship of resources.

Schools

Municipal governments can have a significant influence on school boards within their areas of jurisdiction. While the academic curriculum may be set by a higher level of government, local school boards could encourage extra-curricular activities that get students out into the environment as much as possible, thus offering a better understanding of, and appreciation for, the natural environment. Field trips to environmentally important local areas could be designed to foster a greater appreciation of the value of such areas and engender a sense of responsibility to protect them.

As is the case in all other areas that call for action to solve the problems that led to the present state of environmental damage, there are millions of actions different municipalities could take to protect and enhance the local environment. There are too many to describe here. The role of individual municipal residents is to participate in these programs as much as possible and support them at the ballot box to ensure that the most environmentally aware municipal councils are installed.

Most municipal waste is easily prevented. Regulation of night-time advertising and exterior building illumination could be a simple first step. A change in street lighting operations would be another. Low-pressure sodium vapour lamps are up to ten times more efficient than incandescent lamps. We should question the continual nighttime use of streetlamps in certain areas where they are now left on from dusk to dawn. With modern sensor-activated and computer-controlled systems we have the capability of doing so much better at eliminating unnecessary light pollution.

Of course, one can always argue that we should simply let the market decide. But that is a dereliction of duty. It tells people and communities that if they are wealthy and feel they can afford to waste resources and pollute the environment, then so be it. As in so many other cases of environmental abuse, this "might is right" approach is unhelpful to the environment, to say the least, and will accelerate the deliberate exhaustion of the planet's resources.

Regional Actions

This is a level of governance with tremendous potential for action to heal our wounded planet. There is scope here for a wide range of initiatives and regional autonomy. Regional governments such as counties, townships and regional districts must sometimes resolve conflicts between lower levels of government, like municipalities, and fit within the broad guidelines of higher levels of government at the same time. The range of actions that could be put in place based on bylaws, strategies, and guidelines at the regional level are quite broad.

Two of the most important areas are land-use planning and transportation.

Land Use Planning and Transportation

Working with the municipalities under its wing, a regional government must plan for the overall use of the land within its jurisdiction. This often includes commercial, industrial, residential, recreational, agricultural, and undeveloped land, and transportation corridors between and across these areas.

The biggest obstacle faced by regional government is the apparent conflict between regional economic growth and the protection of its environment. This is an "apparent" conflict because governments at all levels are locked into the old mentality of the last few hundred years; a mentality that flows from the concept that growth is an increase in consumption of stuff rather than an improvement of the quality of life without growth in consumption. The model of economic growth through quality-of-life improvements is more difficult to implement. It is essential that this model be implemented, however, and the old model of consumption laid to rest, if we are to leave a habitable planet for future generations.

Under the old model, growth is most easily achieved by welcoming new arrivals in a region. The new arrivals may be people from other regions of the country, immigrants from abroad, or excess children in existing families of four or more. With more people there comes a need for more homes, more cars, more food, more consumer goods, more production, more jobs, more money, more travel abroad, more this, more that, more...

Each generation of politicians can then look back over their achievements. How many new houses were built when they were in power! How many highways and office towers! How many new malls and factories! How much money flowed into their region! How much more was consumed! How much more was collected in taxes! How much more...!

I live in a relatively small community of about 12,000 people. It is in a region of about 400,000 people at the south end of Vancouver Island, off the west coast of Canada. The entire island, about 450 km long and 100 km wide has a population of about 870,000 people. Mostly forested, the agricultural areas on the island are already insufficient to meet the needs of the people now living on the island. Much food, and almost all manufactured goods must be shipped in from the mainland. For years, I served on the Community Planning Commission in my municipality. This commission was charged with deliberating on many issues that might affect development, and then providing its collective view to the municipal council for consideration. A few years ago, the local regional government developed its Regional Growth Strategy and passed this strategy to all municipalities within the region for their approval. The local municipality asked our Community Planning Commission for comment.

The most striking thing about the Regional Growth Strategy was that it projected a growth of almost 100,000 people, on an island already incapable of sustaining itself without external inputs, in an already congested area with inadequate roads and other infrastructure. The Commission asked how the Regional Government arrived at the projected 100,000-person growth over seventeen years for a region that had a population of fewer than 347,000 people at the beginning of the planning period. The response was that the projection was in line with projections made by other regional governments across the country and it had been put in the drafts from the beginning. The public had not been consulted with respect to the size of the proposed growth.

This is the problem. It is a problem most acute at the regional level, but also a problem at higher levels of governance, right up to international organizations. The first line of defence should be for regional governments to stand by their planning guidelines and balance regional needs for food, water, commerce, and accommodation to the maximum extent possible. Relying on outside sources for basic needs, or asking people within the region to sacrifice agricultural, natural and recreational lands for increased growth in population and commerce is, ultimately, a path to ruin.

Waste Management

Regional governing bodies have jurisdiction over many activities that affect the environment. The reduction, reuse, and recycling of wastes is one such activity. Single-use plastics should be eliminated, reuse of materials should be maximized with the help of education and information programs, recyclable goods should be recycled to the greatest extent possible, and not shipped off to the cheapest offshore disposer. Waste disposal sites should be operated to the highest standards possible and rehabilitated when closed. These activities are often within the purview of regional governments. Regional governments should determine the cost of collection, sorting, recycling, reuse, or disposal of those things which were made to be thrown away, and pass that information to higher levels of government for cost recovery programs that charge those who create waste materials the full cost of properly dealing with their wastes.

Regional governments should provide a wealth of information to their constituents on environmental issues such as climate action, composting at home, drinking water quality, green storm water infrastructure, natural gardening, alternatives to pesticide use, environmental activities at schools in the region, storm water, wastewater and septic discharges, and water conservation.

Economic Inequality

While this book is not focused on social and economic inequality and how these issues should be addressed, it is concerned with the environmental effects of such inequality. Economic inequality often translates into things like multiple homes and large yachts for the wealthy. All such activities should be examined to assess their environmental effects and an appropriate environmental levy applied; a serious enough levy so that the very rich might be encouraged to think that if their money is going to go to environmental causes, they might as well direct it there themselves rather than spending it on frivolous environmentally damaging activities. If they choose to continue environmentally irresponsible activity, within the law, then the revenue collected as the penalty for such misbehaviour should be earmarked and directed by regional governments to the areas most in need of attention for environmental repair.

Provincial Action

In Canada, the aggregation of regions, townships and counties in a geographically defined area is called a Province. In the United States, such an aggregation is a State. Germany has Federal States known as Lander. England has twelve Regions encompassing forty-eight counties and France also uses Regions which, while they do not hold legislative power and thus cannot write their own laws, do have a great deal of discretionary power over infrastructure, public transit, and so on. China also has a Provincial sub-structure, although there is little chance for independent action in their one-party communist form of government. To represent this level of government the term "Provinces" will be used.

Provinces have the authority and the means to implement many of the most significant measures to preserve, protect, and restore the environment. They can establish provincial parks, preserve agricultural lands, set air quality standards and vehicle emission levels. Provinces can establish minimum building standards to conserve energy and offer incentives for builders to go beyond the minimum. They can set standards for landfill operations and prices for the disposal of greenhouse gases into the air. They can change the language of the propagandists for the carbon industry to help carbon consumers understand that they must pay for the emission of pollutants, CO_2 and others, into the air, in the same way that they know they must pay for other waste disposal activities like landfills and sewage treatment plants. Provinces have an opportunity to educate their citizens about the importance of environmental preservation and protection. In Canada and the USA this level of government has the greatest control over school curricula.

Provincial governments can regulate the use of pesticides for both its farmers and the public. They can establish rules for the assessment of new initiatives that might affect the environment to minimize negative effects. Provinces can pass laws with respect to the transportation of dangerous goods, the discharge of effluents from industrial activities, and proper mining operations.

Provinces should establish environmental assessment boards with the authority to investigate the likely environmental effects of major projects intended to address a problem or take advantage of a particular opportunity. In my years as Vice Chairman of the Ontario Environmental Assessment Board (EAB), I saw how effective such quasi-judicial bodies could be as they heard expert evidence on the environmental effects of major projects. Such evidence was cross-examined and re-examined and at the end of the process, the Board assessed all the evidence and issued a decision on whether projects should be denied because they were ill-conceived from an environmental point of view, or should be allowed to proceed, with conditions.

One of the most complex applications I adjudicated was for the establishment of a hazardous waste treatment facility to treat wastes from across the province. That proposal was denied because of the proponent's failure to choose wisely from the alternatives available and select the one least damaging to the environment.

The Halton Region's Landfill application for a very large landfill to serve the entire Halton Region in Ontario demonstrates the power of an effective environmental assessment process. The proponent considered several possible sites, the first few of which they felt were all acceptable from an environmental point of view, and then ranked those acceptable sites in order of suitability. We heard evidence, which was cross-examined thoroughly, from all sides, for each site on the "acceptable" list. At the end of the day, we chose the site which had initially been thought to be the *second* most "acceptable" site as the best alternative from an environmental perspective.

EAB hearings were completely free of outside interference. While the decisions of the Environmental Assessment Board could be appealed to the Provincial Cabinet, and several were, none were overturned during my time there.

For an illustration of the power of provincial or state governments one just has to look to the State of California in the USA. This government is viewed by many as not only a local leader in environmental protection, but also a world leader. It passed a Global Warming Solutions act in 2006 to reduce its global greenhouse gas emissions to the levels they were in 1990. California has already met that target and is planning to reduce such emissions by another 40% by 2030. In support of this, and for other good reasons, California established a Renewables Portfolio Standard in 2002 requiring that 20% of retail sales of electricity would have to come from renewable energy sources by 2010. The target was increased in 2008 to make it 33% of such sales by 2020 and raised again in 2015 to deliver 50% of electricity retail sales from renewables by 2030.

Apart from carbon reduction initiatives, provincial levels of governments can do much in other areas as well. Regulations with respect to the over-packaging of food and other consumer goods, as well as single-use plastics at grocery stores, fast food outlets, and shopping malls can be developed at the provincial level, and thus provide a basis for regional governments and municipalities to act. In many jurisdictions, municipalities are prevented from taking independent environmental action because the province has jurisdiction, not the municipality. That is why provincial governments must get out in front of such issues, to prevent vested interests from damaging local environments when municipalities lack the jurisdiction to protect themselves.

Initiatives at the provincial level often come under attack from parties with vested interests, both political and commercial. A provincial government may put in place a carbon disposal fee only to have a subsequent government with a different agenda come in and eliminate the so-called "carbon tax." This illustrates more than just the normal to-and-fro of politics. It illustrates how powerful interests fund the election of governments, who will protect them from those who would move the world away from the old carbon-based economy towards a new one of carbon equilibrium. Imagine that blacksmiths funded political parties at the turn of the previous century (1900) and persuaded politicians to forbid automobiles because failure to do so would put all blacksmiths out of work. Today's argument is that turning to renewable energy or nuclear energy may put many of the world's coal miners out of work; there are some very rich and powerful people who don't want to see that happen.

The answer is for provincial politicians to see beyond the immediate profit picture of the few to the long-term societal costs of failing to address environmental problems. We can have a few coal and oil magnates now and defer the staggering costs of environmental remediation to generations a century down the road or, we can begin to make changes in our lives now so that we begin to heal this wounded planet before it is too late.

The choice is yours. If you are so preoccupied with having more money in your pocket now, with no concern for the plight of your descendants, go ahead and vote for those who will rescind the "carbon tax." If you are concerned with the survival of future generations, exercise your right to make things better at the ballot box. Just as a single vote has greater weight at the municipal level than at the provincial level, your vote at the provincial level carries more weight than it does at the federal level. If enough provinces in enough countries are persuaded to choose an environmentally sound future, it will be harder for federal level of government to resist.

National Initiatives

National governments can do more than any other level of government to preserve, protect, and improve environmental quality. National programs to protect air, water, and land are under their jurisdiction and control. Their actions can arise from initiatives within appropriate federal departments, be triggered by public demand for action in a given area, or be the result of encouragement from international organizations.

In Canada, several federal departments have responsibilities for activities that affect the environment. Most important are the Ministries responsible for Environment and Climate Change, Fisheries and Oceans, Transport, Natural Resources, and Indigenous and Northern Affairs.

There are also several other agencies whose work directly affects the quality of the environment and federal efforts to protect it. Agencies like the Climate Change Secretariat, the Offshore Petroleum Boards, the Environmental Assessment Agency, and Parks Canada, to name but a few of the eighty-seven such agencies, bureaux, boards, tribunals, and so on.

Within these Ministries, departments, and agencies there are many dedicated and hard-working individuals whose job is to think about threats to the environment, ascertain whether such threats are significant, and propose to their political masters solutions to mitigate or eliminate such threats. To perform their work effectively, they must not be censored by politicians who accord weight to short term economic benefits rather than the survivability of future generations. I have seen too many instances of such behaviour.

I recall such an instance in the early 1970s when mercury contamination levels were reaching, and in some cases surpassing, the levels that had been set for Great Lakes fish. I briefed the minister responsible for fisheries and the environment and suggested that it might be necessary to close the fishery. His reaction was that such action would be too hard on the livelihood of commercial fishermen. He instructed that the regulations be changed to raise the mercury contamination limit tenfold and spare the fishermen.

Not long after this incident one of my scientists noticed that there were reproductive failures and other genetic abnormalities in the Great Lakes Herring Gull population. The prime pollutant suspected in these abnormalities was Polychlorinated Biphenyls (PCBs). The scientist proceeded to study the buildup of PCBs in the Great Lakes ecosystem and its food chain. He found that the concentration increased at every level up to the Herring Gull which was at the top of the food chain within the non-human population. (Humans are at the top of the food chain overall.) So, he decided to see if the contaminants building up in the Great Lakes ecosystem were being carried through to the human population. He undertook a "shopping cart" survey which analysed the contents of a typical basket of food, to determine the levels of environmental contaminants. Shortly after he presented a summary of his work at a local scientific conference, he was ordered by those at the very top of the bureaucracy to not speak on the subject again, nor publish a paper on it, because "the matter of contaminants in food is for the Food and Agriculture Department to determine and is outside the jurisdiction of the Department of Environment." Such censorship is inexcusable.

Not long thereafter there was a release of perchloroethylene from a factory on the shores of the St. Clair River. The perchloroethylene was contaminated with toxic chemicals. I sent divers to investigate and make a video recording of the situation on the river bottom. As soon as I saw the video, I sent a copy of it and an urgent letter to the Federal Minister of the Environment. I said that if this pool of chemicals was in a farmer's field far from drinking water, we would order that it be cleaned up and disposed of as a matter of urgency. Finding it on the banks of a river would elicit an order that it to be picked up immediately. Seeing it in the bottom of a river that supplied drinking water to millions of people downstream suggested that the situation was critical, and we had to act now. The film and letter were couriered to Ottawa for the Minister's attention.

Days passed. I received a call from the Deputy Minister asking if I still had the letter. If so, I was told to destroy the letter and any copies. "You can always discuss these things with me by phone," she said, "No need to send a letter." "What is the Minister going to do?" I asked. "Don't you worry," she replied, "He'll take care of it." More days passed. While I recognized a responsibility to do the Minister's bidding, I was conflicted because I felt that it was also my responsibility to protect the Canadian environment for the Canadian people. Since part of my job at the time was to work directly with the Provincial Minister of Environment, I called the Ontario Environment Minister and briefly described the problem. He asked that we meet at my home immediately to see the film of the toxic pollutants in the river. After watching the film with him, I told him what action was needed to deal with these toxins, and he began to implement that action within hours of our meeting.

The final example of this Federal preoccupation with economic growth pre-empting environmental action came a year later. At a meeting in Washington the Canadian Minister of Environment and the United States Environmental Protection Agency (USEPA) Administrator met to discuss, among other things, the development of a program to deal with toxic chemicals leaking into the Great Lakes. Of highest concern was the contamination of the Niagara River with toxic chemicals from industries primarily on the American side of the river. Another serious concern was the release of toxins from the Love Canal waste disposal site and the Hyde Park and 102nd Street landfills. The Minister instructed me, as Regional Director General for Environment Canada in Ontario, to work with my counterpart, the Regional Director for the EPA's Region II in New York State, to devise a work plan for the clean-up of the Niagara River. I described the events that followed in some detail at the beginning of the chapter about water on page 35.

These examples all demonstrate a shortcoming too prevalent in national governments: they seem to be unable to do what is best for the long-term protection of the planet if they perceive that such action might jeopardize short term political priorities, inevitably economic priorities that are being promoted by special interest lobby groups which fund the politicians who want to be re-elected. If we are to heal this wounded planet, national governments are going to have to think long term, both environmental long term and economic long term. Short term economic gains at the expense of the environment are fleeting, as they will be more than offset by long term environmental damage and economic losses.

Political and business timelines focus on the few years to the next election or the few months to the next annual report. If profits are up, they think all is good. If the GDP went up, politicians expect to get re-elected. Business leaders of tomorrow will find a way to profit from the damage being done now. Perhaps they may profit by investing in novel dike-building technology. Inflatable boats might be sold to those in coastal areas prone to new flooding. Temporary shelter accommodation for those displaced by major storm events could be fabricated and sold to the municipalities likely to be affected. The ultimate proposal to profit from the destruction of the planet: invest in preparatory flights for travel to the nearest habitable planet, a mere 200,000 years away by the fastest interstellar rocket to date. Our objective must not be to climb aboard a rocket ship, hoping to find another planet out there that we might ruin, leaving a desolate planet earth in our wake. We now have the knowledge, tools, and resources to save *this* planet. All we need is the political will, first and foremost the national political will, to do what is necessary.

The possibilities to profit from environmental calamities could be numerous. Businesses do not feel it is their responsibility to protect the environment in a major way. If national governments are to do so, they will have to regulate business in major ways. If there is ever to be a new environmental-political-economic order, the people who elect federal governments will have to speak up with a loud and clear voice. If most electors want their governments to act in favour of the environment, then such governments will be elected.

The list of things national governments might change or regulate is very long. For a start, they might stop subsidizing polluting industries, particularly the fossil fuel industry. In Canada alone, these subsidies are in the two or three billion dollar range each year. Estimates for the United States range from $50 billion to $500 billion a year, depending on how one accounts for the cost of health and environmental effects of burning carbon. Globally, they amount to five trillion dollars. The numbers are less important here than the concept. The idea of taxpayers unwittingly funding an environmentally damaging industry is disgusting. The fossil fuel lobbyists are wealthy and powerful. Here too, if most electors want their governments to make the fossil fuel industry stand on its own, then such governments will be elected.

Countries can take different approaches. Ireland provides an example of how effective a national government can be. The Irish used to get most of their energy from burning peat. It was readily available, cheap, and easy to dig up. Unfortunately, peat is 23% dirtier than coal in terms of carbon emissions. Now, Ireland is down to just a 6% reliance on peat for energy and is on a path to eliminating peat as a fuel by 2030. It plans to replace peat with energy from wind farms. Ireland expects to produce enough of this renewable energy to power the entire island, with some left over to sell to the rest of Europe. While wind power may be unpredictable, tying into the European electricity grid is expected to solve that problem. The important thing here is that Ireland looked for an environmentally friendly solution and found one. A small country, its solution will not solve the world's carbon problems, but its action is a positive example for others.

National governments can, and should, regulate the packaging industry. A good start here might be to require that all producers of packaging materials bear the full cost of complete and final recycling of their products into new materials. The plastics industry produces a million plastic water bottles every minute and sells them to a food or drinks producer. The manufacturer assumes no cost for final disposal of its bottles. The wholesaler pays no fee for packaging its goods in throw-away containers, it pays only the minimal up-front plastic package cost. Similarly, the retailer passes the cost of plastic packaging disposal down the line. The consumer is not aware of the full cost of such packaging either. People buy plastic clamshell-encased tomatoes and bottles of water for only the cost of the basic plastic and its contents. Nowhere is it made clear that all of society will pay for the environmental damages associated with the manufacture of the plastic in the first place, and the local taxpayer will somehow pay for whatever waste recycling and ultimate disposal cost arise because of the packaging.

The cost of the full life cycle of plastic containers should be imposed at the front end of the process. National governments should calculate the environmental costs to society as a whole for the production, handling, distribution, collection, sorting, and recycling or disposal of single-use plastic containers. An appropriate fee should then be imposed on the plastic producer. Thus, the cost to the food or drinks producer would be included in the packaged product. That cost would be reflected in the cost of the finished food package at the wholesaler, retailer, and consumer level. Funds collected from plastics producers should then be earmarked as funds for waste treatment and passed on to the municipalities and regional governments responsible for the ultimate collection and recycling of plastic wastes. Lower-level governments should maintain a separate account for such funds, where expenditures for environmental protection could be seen to match the fees collected from those creating the environmental stress.

If consumers were given the option of buying expensive tomatoes in a clam-shell package or less expensive ones in a paper bag they could then make an informed choice to save the environment and their pocketbook at the same time. They might decide to take a container of water from home rather than pay a much higher price for bottled water that incorporates all the costs involved to ensure that the discarded container doesn't damage the environment. There are about a million plastic water bottles sold around the world each minute; if the proper cost of protecting the environment was incorporated in the price of the product, people who do not use such bottles would no longer be subsidizing those who do.

Similarly, national governments should calculate the expected costs of having to adapt to environment damage from fossil fuel emissions and make that information clearly and broadly available to the people they represent. They should calculate, as accurately as possible, the cost of environmental damages associated with such emissions and charge a carbon disposal fee to every person who discharges carbon combustion products into the air. National governments must stop subsidising the fossil fuel industry, which they did to the tune of $5.2 *trillion* in 2017, for example. Funds generated from carbon disposal fees should be returned to citizens as subsidies for those who choose cleaner energy sources; those subsidies previously directed to fossil-fuel should be transferred to environment-saving activities. This approach uses market forces as a positive element for the healing of the damaged environment. The present situation is one where costs associated with polluting activities by some people are imposed upon everybody. If one person chooses to drive a huge gasoline-powered machine to the corner store, why should the shopper on the bicycle pay for the environmental damage done by the driver of the big SUV? The same approach should apply to industries that emit large quantities of greenhouse gases. The cost of concrete, for example, should include the substantial costs expected to mitigate environmental damages from emissions associated with its production. People could better choose concrete, or a less expensive alternative, if the environmental costs were priced in.

If national governments choose this option, then it should be made clear that this is a fee for CO_2 discharges into everybody's air, not a "carbon tax." People expect to pay a dumping fee when they dispose of an old sofa, not a "someday-to-be-worn-out-and-dumped old sofa tax." They expect to pay a sewage treatment fee, not a "cabbage fibre tax" because much of the fiber in a head of cabbage is going to wind up in a sewage treatment plant. Similarly, paying a CO_2 atmospheric dumping fee is not paying a "carbon tax." The carbon itself is not the problem; the disposal of burnt carbon is.

To be clear here, the costs of mitigating environmental damage from polluting activities are not just the costs of cleaning up after more frequent hurricanes and floods. The costs need to include increased health care costs that arise because the air we breathe is so polluted. Putting all the remediation costs into the up-front price of consumer goods would result in market forces helping consumers to make intelligent choices in favour of the environment and the economy. This is, first and foremost, a national responsibility.

In both the packaging and carbon discharge fee examples described above, the concept is that people ought not to have to bear the hidden costs of corporate activities that damage the environment. Every industry that does something that will ultimately damage the environment for profit should have the cost of preventing or repairing that damage built into the price of the product or activity.

National governments must impose "right to repair" regulations so that consumers are not forced to throw out items that ought to be repairable. We cannot go on passing the final costs of environmental damage down to the last taxpayer. Consumers cannot act responsibly to protect the environment if they do not know the environmental cost of goods and services they consume. Making more damaging goods more expensive than less damaging ones will help consumers choose more environmentally acceptable items.

The same sort of approach would be useful in all areas of human consumption. We consume far more than needed for a satisfying, healthy, high quality of life. The costs of excessive consumption are borne by the environment upon which we all depend for survival. Value Added Taxes and Goods and Services Taxes on non-essential items are seen as consumption taxes in many countries. Using revenues from those taxes for environmental repair in a fully-transparent way may help remind purchasers that there are other costs associated with an excessive lifestyle. Here again, market forces would be used in favour of the environment instead of against it.

A huge problem for national governments is the problem of living within their means in an environmental sense, both nationally and globally. More than any other level of government, national governments are driven by the desire to continually grow their Gross National Product (GNP). It is called Gross *National* Product, after all. This growth is always calculated in material terms, not in quality-of-life terms. The idea of living within a national and global equilibrium framework with respect to population, resource use, and environmental preservation is not part of the political mentality at the national level. Taking positive steps in this area will be the most difficult; it is also the most critical and most important.

One easy first step to reduce unnecessary expenditures would be to stop wasting money on competitive space races. Who can put the first man on the moon? On Mars? Billions of dollars are wasted so people can listen to the wind on Mars for the first time (the Perseverance Mars Rover program). How could those billions have helped save the environment, instead of contributing to its degradation? National governments need to look at the big picture of how their nations fit in a world of finite resources, and the limited ability of the environmental commons to absorb the waste produced while individual nations prosper.

Imagine each nation as an island. Let's say one of these is Island X. Island X is of finite size with finite resources. It has a starting population that has always been in equilibrium with its resources. It has been restrained by health and reproduction limitations. People have died of disease, shortage of food, and poor health, in general. Primitive tools have kept agricultural output low, and absence of technology has prevented the fabrication of sophisticated goods. Those limitations are, in fact, the main reason that island state lived in equilibrium with its environment.

Now imagine that island society learned how to overcome its limitations. It developed new tools and learned how to live longer. They find a comfortable equilibrium with the natural resources that sustain them. They have a high quality of life with few "wealth" assets. At some point, the leaders of this island nation feel that their people will re-elect them only if they can give them more material wealth. They must show them that their economy is growing, year after year. The easiest way to do this is to let the population grow. More children means more children's clothing, more schools, more hospitals, more factories, and more homes. This looks pretty good for the politicians of the day. Island productivity goes up year after year. Eventually the resources of Island X are depleted. All forests that can be cut down to make room for agriculture have been destroyed. All minerals that can be extracted have been extracted. People can no longer live in equilibrium with their environment. They die in increasing numbers because of poor environmental quality, inadequate water, food, and space needed for good health. Island politicians finally realise that the priorities of their predecessors were wrong. Unfortunately, by then too many have suffered and died unnecessarily, because of prior bad policies.

A new batch of politicians realizes that the only path to survival is through improvements in the quality of life rather than increasing the population and the quantity of goods produced and consumed.

The next step towards globalization in this scenario is to imagine that the people of Island X develop new modes of transportation that allow them to trade with a nearby island; Island Y. Island Y is, developmentally, back where Island X was centuries earlier. The visitors from Island X see a great opportunity in Island Y. Here there are new forests, new minerals, potential consumers, and so on. The people of island X bring medicines and training to create a reasonably healthy new workforce. Island Y is then fully exploited. Eventually the resources of Island Y are depleted. All forests that can be cut down to make room for agriculture have been destroyed. All minerals that can be extracted have been extracted. People can no longer live in equilibrium with their environment. They die in increasing numbers because of poor environmental quality, inadequate water, food, and space needed for good health. Island politicians finally realise that the priorities of their predecessors were wrong. Unfortunately, by then too many have suffered and died unnecessarily, because of prior bad policies.

Does this sound familiar? It should. That is the same mistake their ancestors made. Those who forget history are doomed to repeat it. Finally, a new batch of politicians realizes that the only path to survival for Island X and Island Y is through improvements in the quality of life rather than increasing the quantity of goods produced and consumed.

You can imagine this continuing around the world as people find new islands and new continents to conquer. In fact, you don't have to imagine it, you just have to read the exploitative history of planet Earth.

The better path would be for each nation to assess its own capabilities to support its people by using its resources in a sustainable way to provide for the highest quality of life for its people. Having a clear assessment of the national environment, resources available, and population trends, each nation could then determine the path to the optimum quality of life without impinging on the quality of life in other nations. That is, each nation would conduct its affairs in such a way as to not negatively affect the global environmental commons. It is conceivable, in this model, that nations could find ways to share the environmental commons for the benefit of all. Such sharing should be done by consensus. Countries might redistribute both resources and people, but such redistribution must be to benefit of the entire environment and all parties involved. The bilateral and multilateral redistribution of natural resources has been going on for centuries, though it has seldom been fair, equal, or by consensus. The redistribution of populations is a far more difficult issue.

In the past, some country had too few people to use their resources fully and sustainably. This was the case centuries ago; it is not prevalent now because global populations have grown in such an unplanned way that both local and global environments have been badly wounded. To alleviate suffering in some parts of the world, it may be possible to plan for the movement of significant numbers of people from areas where there are now too many, to areas that can still reasonably absorb them. Such migration should only be allowed if the countries from which the migrants arise agree to assess environmentally sustainable population limits within their own boundaries for the future well-being of those countries. Once these limits are determined, national governments need to take steps to see that they are not exceeded. Migration of surplus populations can no longer be used as the final safety valve for jurisdictions in which irresponsible reproduction is still the norm. National targets for populations living in a sustainable global environment may be arrived at with the help of international organizations. Achieving those targets can be helped by individuals and communities around the world donating to international aid efforts for the education and empowerment of girls along with clear information on family planning to achieve the equilibrium needed in each country.

National governments must divert taxpayer dollars from defence spending into foreign aid and diplomatic activities. Defence spending is, by its very nature, spending on things that that are designed to be wasted—wasted bullets, artillery shells, bombs, and rockets made to kill others and destroy their homes and industries with their incalculable associated environmental damage. Much of such expenditures result in the deployment of weaponry in foreign countries under the guise of "protecting national interests" in such countries. All too frequently the interests being protected are commercial ones that have little to do with the security of the country indulging in foreign aggression. Sometimes the aggressor stimulates violent reactions that are felt back in its home country. It is not unreasonable to conclude that the protection of American interests in hydrocarbon extraction throughout the Middle

East led to a perception of persecution in that region; that perceived persecution led to the attacks on the New York World Trade Center; which in turn led to attacks on Afghanistan and Iraq; and that led to counter attacks by Iran throughout the area... Surely mankind is now smart enough to negotiate peaceful settlements to disputes.

Channelling a significant portion of what is now spent on weapons of destruction into life-sustaining foreign aid would be supportive of negotiated settlement of disputes. Putting some, if not most, of the people who now work for the military (more than three million military personnel and contractors in the US alone) to work on positive foreign aid that nurtures the environment would be a good start. With well-directed support to the poorest regions of the world, the will to go to war with one's neighbours would decline. As the discrepancy between the have and have-not nations declines, so also the desire to seize a neighbour's wealth should decline. Add in some thoughtful diplomacy and the result would be a healthier environment in a more peaceful world.

When I began my career in environmental work fifty years ago, there was very little Canadian legislation dealing with environmental issues. The Canada Water Act did exist, so when our research showed that discharges of phosphorus were damaging the Great Lakes environment, it was fairly straightforward to make regulations under that Act, both on the input side, by regulating the removal of Sodium Tripolyphosphate (STP) from laundry detergents, and on the output side, by requiring phosphate removal systems at sewage treatment plants.

But there was no regulation, for example, with respect to dumping wastes in the ocean. Using an international convention against the dumping of wastes at sea, the London Dumping Convention, we were able to develop Canadian legislation prohibiting such activities in waters under Canadian jurisdiction. When passed, this Bill became the Canadian Ocean Dumping Control Act in 1975. That act was extended and strengthened, becoming part of the Canadian Environmental Protection Act in 1999. Similarly, another international convention, concerned with the prevention of pollution from ships, MARPOL 73, led to the development of corresponding Canadian legislation. That legislation was then strengthened by further international work, the 1978 Protocol to the 1973 Convention, developed to better address pollution from oil tankers.

More national environmental legislation was developed to enable Canada to comply with yet another international convention, the Law of the Sea Convention. In that convention, environmental protection articles were developed by a sub-group working on articles for Marine Environmental Protection and the Transfer of Marine Technology.

Both Canada and the United States developed national legislation and regulations to protect their shared environment in the Great Lakes. The Great Lakes Water Quality Board, for which I was Canadian Chairman in the early 1980s, was thus able to set forth sensible and effective steps to protect the Great Lakes environment.

Despite intense lobbying by the fossil-fuel industry, forty-two national governments who are members of the Organization for Economic Cooperation and Development (OECD) and Group of Twenty (G20) countries, had introduced carbon taxes or carbon cap-and-trade programs by 2018. These covered about 15% of global carbon emissions. Just eight years earlier only 4% of such emissions were covered. It is evident that national actions to heal the planet's environment can be stimulated by international consensus and conventions. The next chapter looks a little more deeply into this area of action. So, the trend is definitely in the right direction.

Much has been accomplished by national governments already. Some of the more important Canadian environmental legislation over the last fifty years includes:

1. Canadian Environmental Protection Act
2. Transportation of Dangerous Goods Act
3. Hazardous Products Act
4. Pest Control Products Act
5. Canadian Environmental Assessment Act
6. Fisheries Act
7. Canada Shipping Act
8. Marine Liability Act
9. Navigable Waters Protection Act
10. Oceans Act
11. Canada National Marine Conservation Areas Act
12. Species at Risk Act
13. Migratory Birds Convention Act
14. Canada National Parks Act
15. Environmental Enforcement Act
16. Environmental Violations Administrative Monetary Penalties Act

International Initiatives

By now it should be clear that the "every man for himself!" mentality will not solve the tragedy of the global environmental commons. "Putting *my* community, region, state, or country first" won't do it either. While positive environmental actions by individual sovereign states would be very helpful, a focus on the well-being of individual nations at the expense of the global environment will be counterproductive. International action is necessary to protect the global environmental commons because selfish local, regional, or national interests could otherwise defeat efforts to save the planet. This is especially the case with respect to air and endangered species issues. Water issues tend to require bilateral attention, except for the protection of the world's oceans, which is a major international issue. Although protecting the land component of the global environmental commons properly fits under national jurisdiction, accommodation will be needed when national land use activities could have negative international implications, as may be the case with the Amazonian rain forest, for example.

An excellent example of how international cooperation can work to heal this wounded planet is the Montreal Protocol on substances that deplete the ozone layer. That protocol was finalized in 1987 and was the first international environmental convention to be ratified by all countries. It is a global agreement to protect the stratospheric ozone layer by phasing out the production and use of ozone-depleting substances. The ozone layer had protected our planet from harmful solar ultraviolet radiation for millions of years. In the 1970s a few scientists, Dr. Sherwood Rowland foremost amongst them, concluded that when chlorofluorocarbons (CFC refrigerants, for example) are released into the atmosphere they slowly drift up to the stratosphere and react with Earth's ozone shield. Dr. Rowland and others concluded that this would destroy the ozone layer. The Montreal Protocol, by protecting the ozone layer, will prevent hundreds of millions of cases of skin cancer, tens of millions of skin cancer deaths and hundreds of millions of eye cataracts. Dr. Rowland's science and personal integrity were attacked by the chemical industries that produced chlorofluorocarbons, in yet another example of the perceived conflict between the economy and the environment. Rather than looking for opportunities to formulate and market refrigerants compatible with the environment, industry chose to attack the science, in hopes of continuing along the lazy, business as usual, path.

Dr. Rowland calmly and quietly stood his ground, and in the end was completely vindicated as the mountain of scientific evidence grew to support his initial scientific findings. In 1995, he and his colleagues received a Nobel prize in Chemistry for their work in this field. The Montreal Protocol, which continues to serve as a model for international collaboration to address global problems would not have been possible without Dr. Rowlands courage and steadfastness in speaking scientific truth to power and translating the implications of his work for public health and welfare. Speaking to the need for action in the face of solid predictions about the looming danger of CFCs in the atmosphere Dr. Rowland said, "What is the use of having developed a science well enough to make predictions if, in the end, all we're willing to do is stand around and wait for them to come true."

The current crucial environmental issue for the global community is global warming arising from increasing levels of "greenhouse gases" in the atmosphere. Here, the scientific evidence is clear. Here also, those with vested interests choose to attack the science in the hope that they may continue their lazy path to greater profits. In an attempt to prevent global warming, the international community met near Paris, France to negotiate language for an agreement on how to deal with greenhouse gas emissions. Wording of the agreement had been negotiated by representatives of 195 states belonging to the United Nations Framework Convention on Climate Change, and was signed by all of them by November of 2019. It was encouraging that 187 of those members ratified that agreement, meaning they agreed to be bound by its terms and conditions.

Unfortunately, in June of 2017, Donald Trump, the President of the United States, announced his intention to withdraw the USA from the agreement. Changes in US policy contrary to the Paris Agreement were introduced immediately. Speaking to the subject, President Trump said:

> "[…] as of today, the United States will cease all implementation of the non-binding Paris Accord and the draconian financial and economic burdens the agreement imposes on our country. […] We have among the most abundant energy reserves on the planet, sufficient to lift millions of America's poorest workers out of poverty. Yet, under this agreement, we are effectively putting these reserves under lock and key, taking away the great wealth of our nation—it's great wealth, it's phenomenal wealth; not so long ago, we had no idea we had such wealth […]"

This is the clearest and most tragic example of national selfishness in the environmental field. As was clear in Part I of this book, "The Tragedy of the Environmental Commons," if nations continue to grab more than their share of the world's resources and, in consuming those resources pollute the environmental commons, the world's environment is doomed. In essays published almost two hundred years ago, William Foster Lloyd put it most succinctly: "It is in the interests of the users of a commons to manage it prudently."

The single country responsible for putting the largest portion of greenhouse gases into the environment over the last two hundred years is the United States of America. In countries where the rule of law applies, it is normal that when someone damages someone else's property the party doing the damage pays some form of compensation to the person whose property was damaged. Things are a lot more complicated in the international sphere, but the principle should be the same: the nation that damages others should expect to pay to repair the damage in proportion to the damage caused. International conventions can provide a framework within which compensation arrangements can be agreed.

The world-wide average of CO_2 emissions, per person, grew from three metric tonnes in 1960 to five metric tonnes in 2011. The United States was still putting out 16.5 metric tonnes per person in 2014, while China and India were emitting 7.5 and 1.7 metric tonnes respectively. So, in the quote above, the US President was saying that the US should be allowed to continue to pollute at three times the global average, more than twice as much per capita as China and ten times the level in India, while demanding that these other countries restrain their output. In withdrawing from the Paris Agreement, the American government informed the world that it refuses to pay anything for its past or present damage to the global commons. Had the US taken the same approach to the Montreal Protocol, the hole in the ozone layer would still be growing and there would be many unnecessary deaths, including deaths of American citizens. If the USA continues to avoid its responsibilities with respect to global warming, there will be many more unnecessary deaths, in America and elsewhere, and incalculable damages from future storms and droughts. For the present, the US will continue to enjoy "the great wealth of our nation—its great wealth, its phenomenal wealth…" only until the bill comes due for the neglect of its own environment, and that of the world at large.

Keeping the atmospheric commons clean and healthy should not be viewed as a negative cost of doing business, but rather a positive investment in the future. There are many examples of how direct savings flow from clean air actions. For example, when the auto industry used to spray-paint automobiles without air pollution controls on their spray facilities, tons of organic solvent drifted away to contaminate the air, water, and land of neighbouring states or countries. In response to complaints about this toxic air pollution, the automobile industry installed systems to capture the paint solvents and recover them for re-use. Huge quantities of volatile chemicals were kept out of the atmosphere, to the economic benefit of the environment, and to the benefit of automobile manufacturers, who could re-use those chemicals.

So, here we see both the strength (the Montreal Protocol) and the weakness (the Paris Agreement) with respect to international attempts to find solutions for our wounded atmosphere. If powerful countries refuse to accept their responsibilities, the planet is on a course to disaster. If less powerful countries refuse to accept their responsibilities because the more powerful refuse, the disaster will arrive sooner.

Fostering international cooperation to protect the world's oceans, and bilateral cooperation to protect the Great Lakes, was a major part of my professional career. For me, this work began in the early 1970s with work on the finalization and ratification of the London Dumping Convention (The "Convention on the Prevention of Marine Pollution by Dumping of Wastes and Other Matter 1972"). Prior to its existence, countries could simply load wastes on barges or ships, travel out beyond the limits of individual states' territorial seas, and dump the wastes into the oceans. Waste materials dumped into the oceans included garbage, chemicals, and even radioactive materials. Back in Canada, I worked with provincial governments to develop the provisions of the Ocean Dumping Control Bill, which eventually became the Canadian Ocean Dumping Control Act. The London Dumping Convention was upgraded in 1996 under the London Protocol, and the Canadian Ocean Dumping Control Act was eventually incorporated into the Canadian Environmental Protection Act in 1999.

In the mid 1970s, the world's oceans were being polluted by very large oil tankers carrying crude oil, primarily from the Gulf States of Saudi Arabia and nearby countries to markets in North America, Europe, and Asia. The pollution was both operational (pumping of ballast water contaminated with crude oil), and accidental (tanker wrecks were common at the time). An international convention was developed and ratified to deal with pollution from ships in general, The International Convention for the Prevention of Pollution from Ships, 1973, or MARPOL 73 (Marine Pollution convention of 1973), as it was commonly called. Unfortunately, it proved inadequate to deal with oil tanker pollution.

The Marine Environmental Protection Committee of the Intergovernmental Consultative Organization, of which I was Chairman in the mid 1970s, was charged with the task of developing specific measures to apply to oil tankers at sea. Some Very Large Crude Carriers (VLCCs) had run aground around the world causing terrible pollution of the marine environment. Other VLCCs were also causing chronic pollution because of their operational practices, like pumping out contaminated ballast water as they approached their loading ports. So, we drafted a wide range of anti-pollution measures. Things like segregated ballast tanks, protective location of ballast tanks, back up radar, and back up steering. (The rudder on the 234,000-ton Amoco Cadiz had jammed and the crew couldn't fix it. It ran aground on the coast of France, causing an immense oil spill.)

One of the more interesting discussions in the Marine Environment Protection Committee was the development of a requirement for Crude Oil Washing of tanks before a tanker left its off-loading port. Light-weight crude oil from the upper levels of a tank adjacent to one which had just been emptied would be pumped through high-pressure nozzles in the just-emptied tank to scour away the "clingage," (the heavy waxy coating of viscous oils that forms on the walls of cargo tanks as tankers move through cold ocean waters), and pump that clingage to the refinery for subsequent use. When the tanker filled those tanks with ballast water for its return voyage, it would be putting that ballast in much cleaner tanks and not discharging contaminated ballast as it approached the loading port.

The tanker industry initially opposed this provision because of the cost of installing the Crude Oil Washing system. After implementation was imposed, in keeping with the provisions of the 1978 Protocol to the 1973 MARPOL Convention, tanker owners and operators proudly portrayed their efforts as a big contribution to cleaning up the world's oceans. In fact, they were more pleased to see that the oil recovered in the Crude Oil Washing process and delivered to refineries quickly paid for the cost of installation and operation of the system. Think about that for a minute; if a 50,000-tonne tanker recovers 1% of cargo through Crude Oil Washing, that is 500 tonnes or about 3,700 barrels of oil. With oil priced at around $60/barrel, that was an additional $222,000 per trip.

Another important provision that arose in the drafting of the 1978 MARPOL Protocol was the provision for enforcement. If a tanker entered the waters of a country party to the convention without the proper measures for pollution prevention, that country could hold that tanker in its port until it was brought into compliance. This was a powerful incentive for all parties to comply. This is the first time, to my knowledge, such an enforcement provision had been included in an international convention. In essence, the "Flag State" in which the tanker was registered ceded control over the tanker to a coastal state, if the tanker did not comply with the rules of the day.

I have described this particular aspect of the international marine pollution regime to draw attention to the fact that often there are economic advantages to solving the world's environmental problems. In the case of Crude Oil Washing, 3,700 barrels of oil which might otherwise have been dumped in the oceans each trip were saved and put to other uses. That's 3,700 barrels less oil that needed to be pumped as part of subsequent trips, and many such trips avoided along with the environmental damage of such trips.

At the highest level of international cooperation, measures for the protection of the marine environment were drafted during negotiations of the Law of the Sea Convention. At these negotiations in the 1970s and 1980s, a group of delegates from several nations drafted specific provisions for the protection of the marine environment. We focused on measures for countries "to prevent, reduce, and control pollution of the marine environment from any source, using for this purpose the best practicable means at their disposal." We also drafted measures by which countries would "ensure that activities under their jurisdiction or control are so conducted as not to cause damage by pollution to other States and their environment;" and measures to ensure "that pollution arising from incidents or activities under their jurisdiction or control does not spread beyond the areas where they exercise sovereign rights." The measures we drafted included the following:

> "...measures taken pursuant to this Part (Part XII of the United Nations Convention on the Law of the Sea) shall deal with all sources of pollution of the marine environment. These measures shall include those designed to minimize to the fullest possible extent:
>
> a. the release of toxic, harmful or noxious substances, especially those which are persistent, from land-based sources, from or through the atmosphere or by dumping;
>
> b. pollution from vessels, in particular, measures for preventing accidents and dealing with emergencies, ensuring the safety of operations at sea, preventing intentional and unintentional discharges, and regulating the design, construction, equipment, operation and manning of vessels.
>
> c. pollution from installations and devices used in exploration or exploitation of the natural resources of the sea-bed and subsoil, in particular measures for preventing accidents and dealing with emergencies, ensuring the safety of operations at sea, and regulating the design, construction, equipment, operation and manning of such installations or devices.
>
> d. pollution from other installations and devices operating in the marine environment, in particular measures for preventing accidents and dealing with emergencies, ensuring the safety of operations at sea, and regulating the design, construction, equipment, operation and manning of such installations or devices."

These international conventions, and national legislation that enables them in countries around the world, are a clear illustration of how to heal the planet's wounded oceans.

There are also international measures to protect the world's water that are not global but are quite critical and often complex. As an example, in the mid-1970s, sixteen countries surrounding the Mediterranean Sea got together and drafted articles to protect the Mediterranean. They accepted these as part of a Mediterranean action plan and, in 1995, formalized them as The Barcelona Convention for the Protection of the Marine Environment and the Coastal Region of the Mediterranean. The convention provides measures to prevent dumping things in the sea, prevent ship discharges, prevent pollution coming from the surrounding land, and to cooperate in ocean science and technology, and data sharing, among other things.

Closer to home, we have the *Northwest Atlantic Fisheries Organization* (NAFO) which replaced the *International Commission for the Northwest Atlantic Fisheries* (ICNAF), which had been put in place in the 1950s. Both organizations intended to ensure the conservation and management of fish stocks in the region, particularly the Atlantic Cod. Members of this organization recognized that something they called "optimum sustainable yield," which is the catching of the most fish with the most efficient use of time and energy, is better than the "maximum sustainable yield," where one applies all the energy and time needed to take the most fish possible while still maintaining the overall population in reasonable health. Maximum sustainable yield was better than "maximum yield," in which one just took all the fish possible and let the species spiral down to extinction.

Self-interest of some member states occasionally defeated efforts to preserve stocks. I recall the angst of my marine biology colleagues at the Law of the Sea negotiations as they recounted how members of ICNAF decided to increase quotas one year because they had been unable to fill their previous year's quota—hardly the way to protect resources. Here we see how selfish interests cause catastrophic environmental damage. Modern fishing technology had enabled the capture of so many fish that "maximum catch" had replaced "maximum sustainable catch" and "optimum sustainable catch" as the operating principle. By 1993, six Atlantic Cod populations had collapsed, forcing a belated moratorium on fishing those stocks.

Protection of the global commons is required in four areas: the atmosphere, water, land, and species. In terms of international action, measures to protect the atmosphere are most critical because the movement of air absolutely knows no boundaries. International measures to protect water are next in importance because for some bodies of water, especially the world's oceans, borders are again meaningless. Protection of the world's land resources is another matter.

The uses of a country's land area is, quite reasonably, under the control of the country that has jurisdiction there. The difficulty here is that some of the world's remaining forested lands are important for helping the world get through the next few decades of global warming and stresses on natural systems that produce oxygen. The composition of Earth's atmosphere is mostly Nitrogen (78%), with Oxygen at 21%, and a few others gases, like Argon and Carbon Dioxide in that last 1%. Trees and other plants "breathe in" carbon dioxide and "breath out" oxygen. Animals breathe in Oxygen, burn it in their cells to get energy, and then exhale Carbon Dioxide. For millennia, this system was in equilibrium: there were enough trees producing enough Oxygen to supply the needs of animals for energy. Now, Oxygen is used for transportation (cars, trucks, airplanes, trains, and boats all burn it up), keeping people warm (oil and gas furnaces), providing electricity in fossil-fueled generating facilities, and is consumed when forests burn.

It is hard to get a good number for CO_2 resulting from deforestation around the world. Almost all deforestation is unregulated. In some cases, trees are cut down and harvested for lumber. Often, they are burned or left to rot to provide space for future land for agriculture. While burning or rotting, they produce CO_2. A reasonable estimate is that ten million hectares are deforested each year, an area larger than New Brunswick (almost the size of Ohio, for American readers), of which almost four million hectares were primary rainforest. Most of the loss of primary rainforest was in Brazil and Indonesia. In addition to deliberate deforestation, there were probably more than ten million hectares of forests lost to wildfires around the world, although accurate figures are impossible to find in this area.

This is where national and international imperatives cross. Brazilians and Indonesians see an economic opportunity in clearing their forests and planting crops much in demand around the world. They have sovereign control over their land. The world community sees the preservation of forests and woodlands as important for maintaining the flow of Oxygen and stemming the flow of CO_2 for the whole world. If Brazil, for example, will not maintain the Amazon rainforest out of goodwill (and in its own environmental interest), and the world community sees environmental value in that rainforest, then an accommodation is needed; an accommodation that recognizes the mutually supportive environmental and economic goals.

It is reasonable that those countries most responsible for the historic and on-going increase of greenhouse gases in the atmosphere should compensate those who agree to sacrifice national economic growth in favour of the global environment. The compensation should be equal to the economic sacrifices made by those countries that agree to protect their forests and do so. Unfortunately, the world's biggest polluters refuse to recognize their responsibilities in this area.

The fourth pillar of the environmental commons is the protection of plant and animal species around the world. Every year there are billions of dollars of trade in endangered plants and animals, as well as consumer goods made from endangered species. These include ordinary things, such as timber and leather, and unusual items like tourist curios and products felt to be of medicinal value. It is so sad to think about the killing of a magnificent animal like the elephant so that its tusks can be cut off and carved into cute trinkets for the very rich. It is equally tragic to see the carcass of a rhinoceros with its horn hacked off because someone can sell ground-up rhinoceros horn in the belief that it will treat certain ailments.

Chemically, rhinoceros horn is like human fingernails or hair. It is unreasonable to think that eating ground-up hair would cure typhoid or hallucinations. There is no evidence that it does. There is no evidence that ground-up Rhino horn will do so either. Unfortunately, more than four centuries ago, a Chinese pharmacist named Li Shi Chen stated that rhino horn could cure snakebite, typhoid, devil possession, hallucinations, and carbuncles, among other things. There are no scientific studies, let alone placebo-controlled, double blind verifiable scientific studies, to support any of these claims.

So, thousands of rhinoceros were killed to accommodate a modern-day fiction that stemmed from an ancient belief. In the modern world, fortunately, rhinoceros, elephants, and almost 39,000 other known species, ranging from Antilocapridae Antilocapra americana to Zygophyllaceae Guaiacum unijugum, now fall within the protective envelope of the Convention on International Trade in Endangered Species of Wild Flora or Fauna (CITES). This convention aims to ensure that the international trade of plants and animals does not threaten their survival. It is believed that the ivory ban that arose from the work of CITES greatly reduced the poaching of African Elephants, not so much because of the listing itself, but rather because people came to understand that ivory trade is harmful and illegal because of the tremendous publicity surrounding the promulgation of the ivory ban.

To feel the cruelty of a world without CITES, try to imagine that another species superior to human beings emerges on this planet. Animals of that species feel that there is some magical medicinal power in our hair and fingernails, and they would like to make trinkets from our teeth. So, they kill every man, woman, and child on the planet to satisfy their cravings. How would you feel about that?

Solutions for a Wounded Planet

I recall so clearly the answer given by the famous primatologist Jane Goodall to a question put by a radio interviewer a few years ago. The question was concerned with to how to decide, if two species were being threatened by human activity, which one should be left to die to save the other, a kind of endangered species triage. There was silence for a few seconds, then Jane said something like: "Imagine if all the animals of the world were able to talk to each other. They have a meeting because they see that the world is under serious threat and one species must be eliminated so the others could survive. Which species would they vote to eliminate to save the planet?" A more prolonged silence ensued; everyone knew the answer, no one spoke.

We have an obligation to heal this wounded planet through international cooperation. That will require all nations to work together. They will only do so if the people, the individuals, families, and communities they represent let their own national governments know that they must act, and act collectively with other nations, to save the environment.

It is sometimes said that the weakness of international agreements and conventions is the absence of an enforcement capability. That is certainly a weakness. There are, however, ways to overcome it. International conventions should include provisions which allow sovereign states to take carefully considered and prescribed actions for enforcement. The 1978 Protocol to the MARPOL marine pollution convention described earlier is a concrete example of how this may be done. Similar measures should be included in all environmental conventions.

The greatest threat to the planet now is global warming from the burning of fossil fuels. The solutions to this deliberate damage include using everything in the international toolbox now to control this, and add new tools as required. For example, the World Trade Organization (WTO) should allow and encourage tariffs on all imports proportional to the extent to which items imported have above-average CO_2 content. If a country exports an item with above average CO_2 impact, appropriate tariffs should be applied. Market forces would support the production of goods with lower impacts.

Those most responsible for historic CO_2 pollution should contribute to an international fund to help those countries most affected by such pollution adapt to the effects.

Some international environmental agreements link only a limited number of countries. These bilateral or sometimes trilateral agreements are only binding for the countries that have ratified them but are nevertheless essential in the international environmental regime. A great deal of progress has been made, and work continues, to solve environmental problems that go beyond national boundaries.

Solutions for this wounded planet will be most difficult and complicated at the international level. If every individual, family, community, municipality, and province has been committed and sincere in their support for recovering from environmental damage and preventing it in the future, their governments will have to respond accordingly. There is hope in this area. More than 3,000 international environmental instruments, regional or global in scope, have been identified. A few of the more important ones are listed here:

- Barcelona Convention for Protection against Pollution in the Mediterranean Sea, 1976
- Convention on the Conservation of Migratory Species of Wild Animals (CMS), Bonn, 1979
- Convention on Fishing and Conservation of Living Resources of the High Seas
- Convention on the International Trade in Endangered Species of Wild Flora and Fauna (CITES), Washington DC, 1973
- Convention on the Prevention of Marine Pollution by Dumping Wastes and Other Matter
- International Convention for the Prevention of Pollution of the Sea by Oil, London, 1954, 1962, 1969
- International Convention for the Prevention of Pollution from Ships, 1973 as modified by the Protocol of 1978 (MARPOL 73/78)
- International Convention for the Regulation of Whaling (ICRW), Washington, 1946
- Kyoto Protocol - greenhouse gas emission reductions
- Montreal Protocol on Substances that Deplete the Ozone Layer, Montreal, 1989
- United Nations Convention on the Law of the Sea
- United Nations Framework Convention on Climate Change
- Vienna Convention for the Protection of the Ozone Layer, Vienna, 1985, including the Montreal Protocol on Substances that Deplete the Ozone Layer, Montreal 1987

Before moving on to the final chapter, you may remember how we started with William Forster Lloyd's *Tragedy of the Commons*. As long as every user of the commons respected the environment and did not abuse it for individual economic gain, all benefitted. A soon as one or more abused the commons for their own benefit, all suffered. The international regime for protecting the global environmental commons works the same way—everyone needs to be a positive player. If some are not, then there need to be proper and meaningful penalties for those who will not play by the rules, as was the case for the 1978 Protocol to the MARPOL Convention with respect to tanker pollution.

Conclusions

SEX, RELIGION AND POLITICS

It is obvious from previous chapters that to heal this planet we need to address the disequilibrium that has emerged between species. There are simply too many people for planet earth to sustain with a high quality of life in a high-quality environment. From this disequilibrium flows the host of problems described earlier, from the misappropriation of land to the contamination of the atmosphere.

Sensible planning of family size is often opposed by certain religions, perhaps to avoid losing their share of adherents, or because they disagree with the idea that women should be empowered to be in control of their own bodies.

Stabilization or reduction of populations is often opposed by business interests because it could reduce consumption. Reducing consumption would hurt the bottom line of present-day businesses. Businesses inevitably support the constant growth of GDP. National governments are sometimes more representative of the businesses that support and lobby them, than the people who elect them, if indeed, they are elected at all. In many nations, religious rulers are also government rulers.

So, here we see the three greatest causes of wounds to the environment: Irresponsible reproduction, Religion, and Politics. The three things I was taught, as a child, to never discuss in polite company.

If we want to heal this planet, it is time to talk about all three.

The fusion of church and state has deep roots, spanning many centuries, supressing women's rights and freedoms along the way. Through those centuries, wellbeing has come to be measured in terms of the wealth of individuals and states rather than the quality of the environment in which we live. Individual wealth has been linked to the wealth of the state. It is easiest to grow state wealth by increasing the size of the market. The size of the market will grow as the population grows. Here, the interests of markets, governments, and some religions are aligned. Growth in population is made easier by keeping women disempowered, and that is most easily achieved by denying education to girls.

It is time to talk about finding a way, at the international, national, and provincial levels to ensure that girls have the same opportunities as boys when it comes to education so that they may become as empowered as men when they grow to be women.

To do that we will need the cooperation of the world's religions.

But we need their cooperation to do more than stop environmental damage from overpopulation. We need their cooperation to stop the disastrous environmental damage that flows from wars and preparations for wars.

In his 1966 work, Swiss Theologian Hans Kung wrote that there can be:

> *"No peace among the nations without peace among the religions.*
> *No peace among the religions without dialogue between the religions*
> *No dialogue between the religions without investigation of the foundation of the religions"*

It is time to investigate the foundations of religions and start the dialogue.

Finally, we must learn that short-term economic gain that wounds the environment leads to an unlivable environment and economic collapse.

Sex, Religion and Politics?

It's time to talk.

EQUILIBRIUM

We live in a world of disastrous disequilibrium. Everyone wants more wealth, more stuff, more travel, more exciting experiences… This disequilibrium is destroying the planet's environment.

To heal this planet's wounded environment, we need to achieve equilibrium. We need a world in which everyone has sufficient food, shelter, health care, education, and room to breathe and enjoy the natural environment.

What needs to be done about this is clear.

People around the world need to let their politicians know that the preservation of a livable environment for present and future generations is *the* political priority. We can begin this process at the local level, demanding that everything possible be done to provide better transportation, more efficient housing, local food production, and so on. It is not that we are killing jobs, rather that we need jobs that save the environment, not destroy it. There would be a comfortable standard of living in a healthier environment. We would have better services instead of more "stuff." People would be employed in more meaningful jobs, at fair wages, to keep the planet healthy.

Provincial or state governments need to hear and respond to the priorities of the populace or be turfed out. What is being put in place at the local level needs to be built upon at the federal level. This takes work, intelligence, and some measure of sacrifice by the present generation. Sacrifice in the sense that we may have to deny ourselves the questionable pleasure of overeating, bigger cars and homes than we need, another trip to a faraway land, and so on. Not great sacrifices to keep the world livable. If local and regional governments do their job, and politicians at the national level come to realize that their electors really want a better quality of life through a restored environment, they may realize that their governments depend on satisfying those desires.

National governments need to do a kind of "whole country environmental assessment." How many people can a country support while maintaining a high quality of life for all? How can the nefarious nexus between religious leaders, population growth, and a political/industrial focus on economic growth flowing from producing more things, be broken? These questions can be answered.

The answers require that every citizen revise expectations to focus on a healthier environment and a higher quality of life. A new equilibrium without growth in the production and consumption of things, in favour of growth in the quality of human services, is a good thing. Here are a few of the many possible solutions to the political malaise that destroys the environment:

1. Demand that politicians assess the environmental effects of every proposal they make.
2. Encourage politicians at every level to assess the carrying capacity of the area over which they have jurisdiction to provide for sustainable development in that area. Simply put, "carrying capacity" is the *maximum* amount of human activity that an area can support with a healthy environment. As was the case in making the distinction between 'maximum sustainable catch' and 'optimum sustainable catch' with respect to fisheries, we should strive for an optimum sustainable carrying capacity, not necessarily the maximum sustainable capacity.

3. Request measures that punish environmentally damaging conspicuous consumption in all areas, from transportation, through housing, to consumer goods.
4. Ask politicians to shift resources from military spending to the repair of the environment.
5. Remove the link between religious authority and political power.

The ultimate solution for this wounded planet is to reject the old idea that a high quality life is a life of 'prosperity', equivalent to having more stuff. In its place we need to embrace a new approach—that true well-being flows from a high quality of life in a healthy environment where humans live in equilibrium with all else on this planet.

We have the knowledge and the means to heal this wounded planet. The only question that remains is: Are we too selfish and lazy to do it?

Why I wrote this book

This book had to be written. I didn't want to write it and, frankly, it wasn't a very pleasant task, but it had to be written. While there have been many excellent books written about various environmental problems, I have not yet found one that knits together the problems of the waste society, the damage to various components of the environment and the "greed" society in which individuals seek to acquire power and garner a disproportionate share of the world's limited resources.

After a lifetime of interest in the environment, and work in the environmental field at the local, national, and international levels, I have witnessed well-meaning individuals and governments nibbling away at the edges of environmental problems while ignoring "the elephant in the room," which is irresponsible reproduction, power hungry religions, and economics-driven governments. These matters are seldom, if ever, discussed in an environmental context and practically never acted upon.

So now that the writing is done, I expect many will pronounce this to be a depressingly pessimistic view of our evolution in relation to our environment. I will not rise to the bait with the usual response that "This is not pessimistic; it is realistic!" Instead, I leave it to you to judge.

If you believe that mankind will consciously reduce its population to a more sustainable level through the education of girls and subsequent family planning at the local, national, and international levels before incurring unnecessary famine, disease, and death; if you believe that the religions of the world will accept that they do not need a larger share of the world's population to support their religious beliefs, and will learn to live in peace and harmony with all other religions; if you believe that those who hold power in corporations and government will accept a decline in material wealth along with the corresponding rapid drop in profits for extractive and wasteful industries and a corresponding decrease in traditional GDP; if you believe that the world will quickly understand and accept that our real hope for prosperity and environmental survival lies in seeking quality over quantity at every opportunity; and, *if you are right*, then I will have to accept your assumption about pessimism here. You will become the realist. If, on the other hand, you come to a different conclusion after serious reflection and thought about these issues, then you may arrive at a different appreciation of realism.

If this book helps us to choose a path to a new equilibrium with our environment, with a significant reduction of unnecessary suffering and death along the way, then it will have been worth the effort of writing it.

That is why I wrote this book, and it is not really such a new idea:

NON NOBIS SOLUM. EX HAC TELLURE MERITA NOSTRA CURAE ET IN OMNE TEMPUS, IN COMMODIS OMNES SPECIES.
(ADAPTED FROM CICERO - 44 BC)

Not for ourselves are we born.
This planet merits our care for all time and for all species.

LIST OF FIGURES AND TABLES

Figures

		Page
1.	The Greenhouse Effect	13
2.	Carbon Moving into the Atmosphere	15
3.	Arctic Ice Cover Over Time	19
4.	Changes in Global Temperatures over the Years	20
5.	Carbon Emissions and Global Temperatures	24
6.	Heat Content of the World's Oceans	27
7.	Personal Share of Fresh Water	38
8.	Personal Share of Earth's Usable Land	46
9.	Personal Share of Usable Land for Living Space	47
10.	Concentration of People in Urban Settings	48
11.	Stacked Populations in Cities	49
12.	Number of Species over Time	54
13.	Sperm Count and Plastic Production	93
14.	CO_2 in the Atmosphere and Annual Emissions	99
15.	Carbon Emissions by Country	101
16.	Per Capita CO_2 Emissions	101
17.	Global Energy Production and Use	105
18.	Grams of CO_2 per Person, per Kilometer	111
19.	International Tourist Airline Trips	113

Tables

1.	Total Available Carbon	14
2.	Endangered Species	56
3.	Critically Endangered Species	57
4.	Canadian Environmental Legislation	171
5.	International Agreements and Conventions	182

CPSIA information can be obtained
at www.ICGtesting.com
Printed in the USA
LVHW012143301022
731915LV00006B/420